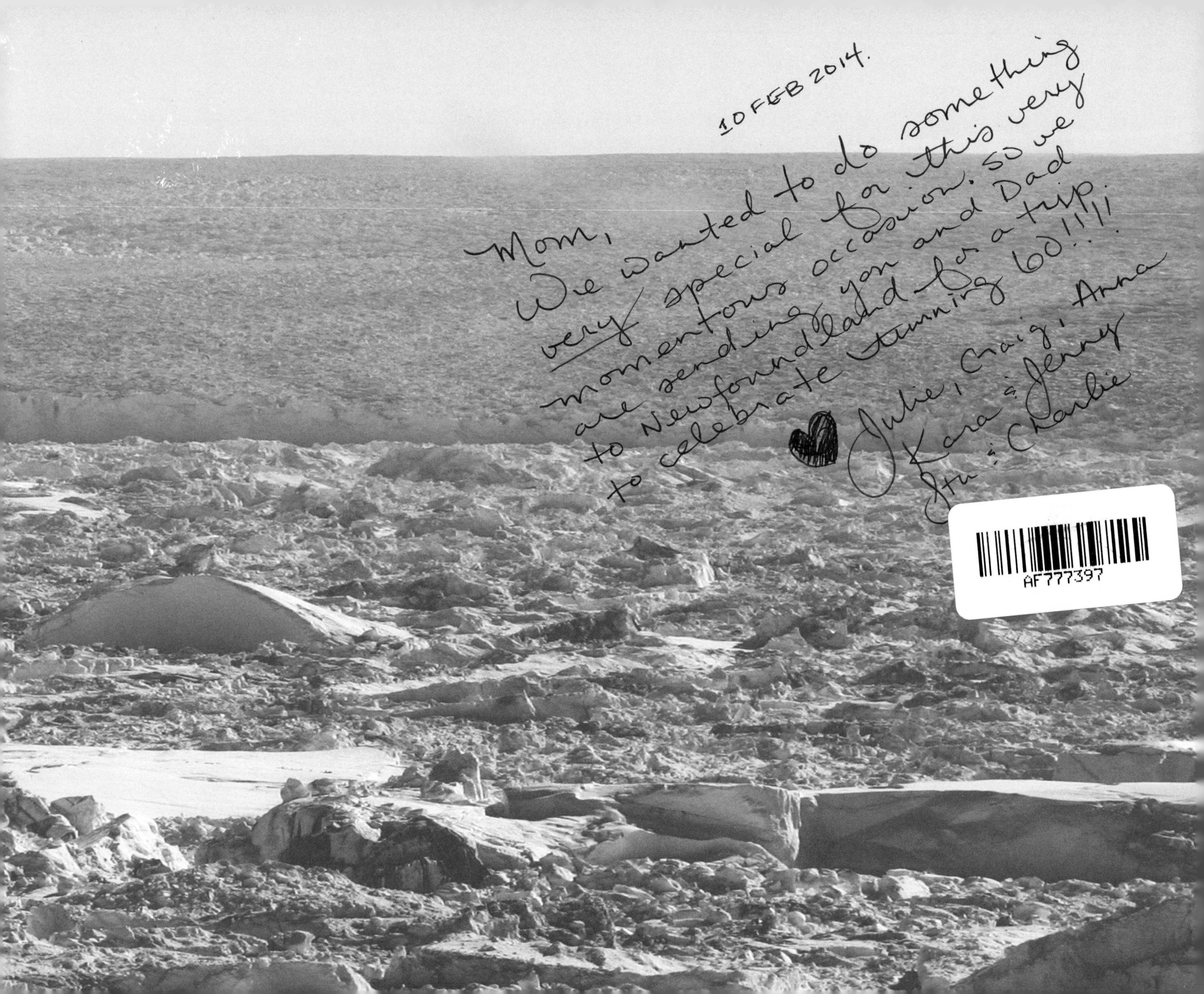

10 FEB 2014.

Mom,
We wanted to do something
very special for this very
momentous occasion. So we
are sending you and Dad
to Newfoundland for a trip
to celebrate turning 60!!!!!

♥ Julie, Craig, Anna
Kara & Jenny
Stu & Charlie

Iceberg Alley

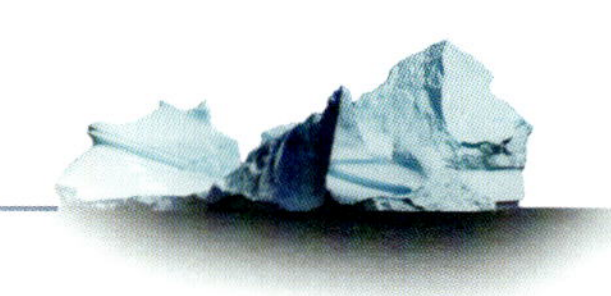

Library and Archives Canada Cataloguing in Publication

Bruneau, Stephen E. (Stephen Earl), 1964-
Iceberg Alley : a journal of nature's most awesome migration / Stephen Bruneau & Kevin Redmond.

ISBN 978-1-897317-71-6

1. Bruneau, Stephen E. (Stephen Earl), 1964-. 2. Redmond, Kevin. 3. Icebergs--Labrador Sea. 4. Icebergs--Labrador Sea--Pictorial works. 5. Icebergs--Davis Strait. 6. Icebergs--Davis Strait--Pictorial works. 7. Icebergs--Newfoundland and Labrador--Atlantic Coast. 8. Icebergs--Newfoundland and Labrador--Atlantic Coast--Pictorial works. I. Redmond, Kevin II. Title.

GB2403.2.B78 2010 551.3'42 C2010-902858-9

Layout: Albert Taylor

PRINTED IN CANADA

Mixed Sources
Cert no. SW-COC-001271
© 1996 FSC
FSC

This text of this book is printed on Ancient Forest Friendly paper, FSC certified, that is chlorine-free and 10% post-consumer waste.

FLANKER PRESS
PO BOX 2522, STATION C
ST. JOHN'S, NL, CANADA
TOLL FREE: 1-866-739-4420
WWW.FLANKERPRESS.COM

15 14 13 12 11 10 1 2 3 4 5 6 7 8 9 10

Canada
Canada Council for the Arts Conseil des Arts du Canada
Newfoundland Labrador

We acknowledge the financial support of the Government of Canada through the Book Publishing Industry Development Program (BPIDP) for our publishing activities; the Canada Council for the Arts which last year invested $20.1 million in writing and publishing throughout Canada; the Government of Newfoundland and Labrador, Department of Tourism, Culture and Recreation.

ILULISSAT • GREENLAND
CAPE SPEAR • NEWFOUNDLAND

IcebergAlley

Stephen Bruneau & Kevin Redmond

A Journal of Nature's Most Awesome Migration

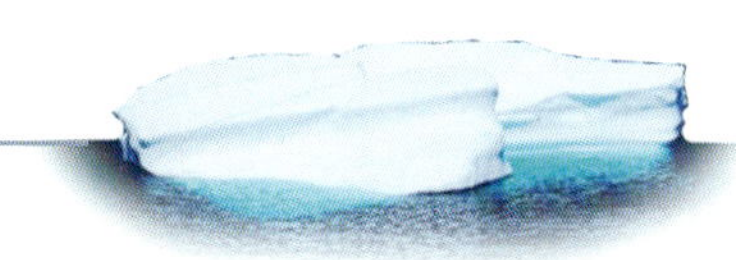

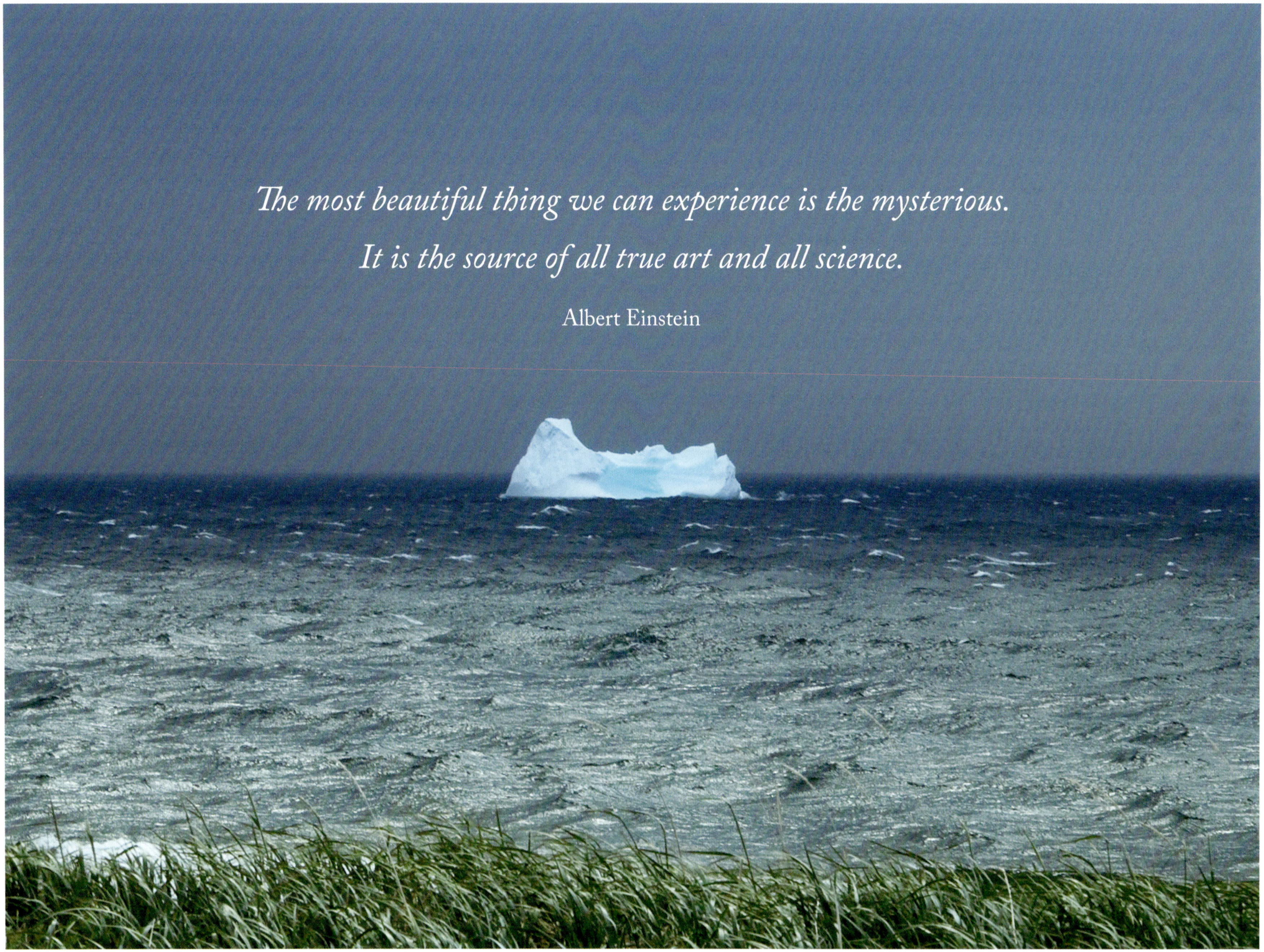
The most beautiful thing we can experience is the mysterious.
It is the source of all true art and all science.
Albert Einstein

Acknowledgements

We would like to thank the many people who have contributed in any way towards the creation of this book. Of special note are the following:

All those kind boaters willing to take us along including: Ed Kean, Chris Kearney, Tom O'Dell, Walter Fowler, Joe Fowler, George Fowler, Ricky Gibbons, Francis Smith, Mike Earle.

To those who shared their local expertise in locating icebergs and welcomed us to their favourite haunts and homes while we visited: Mary Foley, Cyril McGrath, Michelle White, Mike Earle, Gordon Slade, Andrea Cote.

To those who have vetted facts and supported us in our research in compiling the manuscript: Angus Bruneau, Dave Fowler, Sarah Roberts, James Youden, Mike Earle and Phil Power the "Iceberg man."

To those who have graced us with their company in our iceberg quest elevating our own personal experience we thank you for sharing this journey with us. Angus Bruneau, Denny Christian, Greg Crocker, Pat Dornan, Dan Murphy, Eric West, Ed Redmond.

To our immediate families: Mary, David & Laura Bruneau; Sophia, Thomas, Susan & Jacquelyn Redmond; your willingness to go along on our many iceberg excursions, patience while we avail of another "photo op" and support as we complete our final drafts and edits is always appreciated and remembered.

To Angus Bruneau for the use of his images found on page twenty-seven and one-hundred-three.

To Robert Collingwood, Dave Hood, Dave Sacks and the Iceberg Vodka Corporation, Joseph Butler, Butler Enterprises Limited; your financial support helped ease our financial burden in exploring icebergs in the remote reaches of Iceberg Alley.

To graphic designer Albert Taylor your expertise has brought quality to the book, your patience and guidance has made this process a pleasurable and learning experience for us both, thank you!

To Garry Cranford, Jerry Cranford and the staff at Flanker, with your interest, support and professional guidance you have made the process pleasurable and the book a product of which we are all proud.

Dedication

Thanks to my father, Angus Bruneau, for his unwavering interest and enthusiasm. Many things would simply not have happened if not for his appetite for exploration and generosity. Thanks also to my mother, Jean, for purchasing most of the copies of my first book.

Steve

Thanks to my mother, Marjorie Redmond, for sharing her passion of our natural world, and my father, Michael Redmond, for sharing his passion of photography. Through their loves, I continue to experience the mysterious, unravelling some of its secrets while it still remains a source of lifelong intrigue, learning, and pleasure.

Kevin

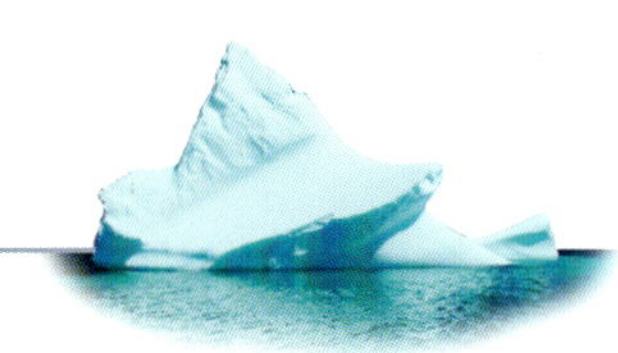

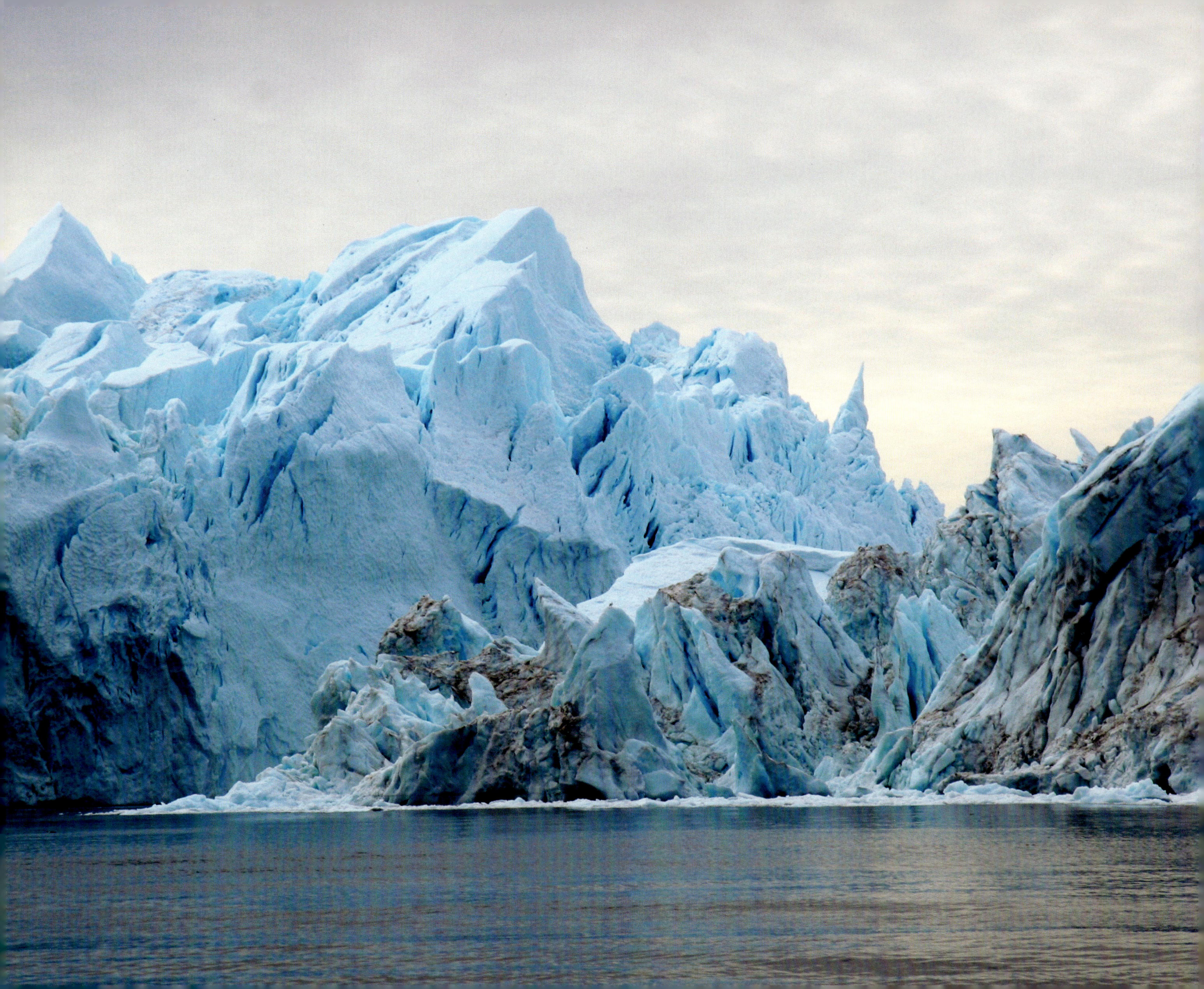

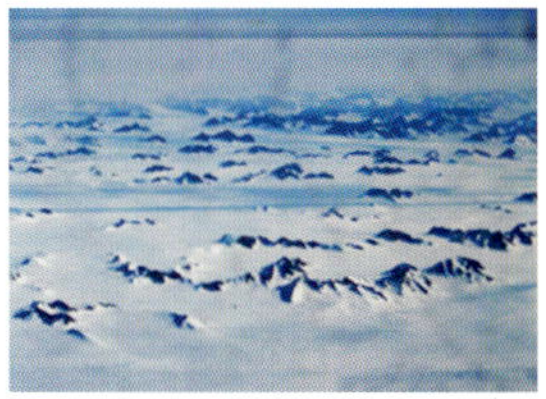

Table of Contents

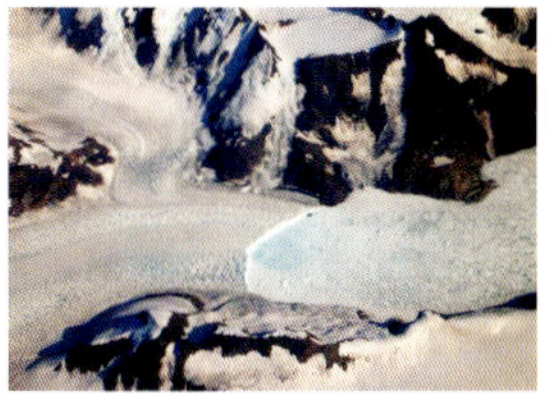
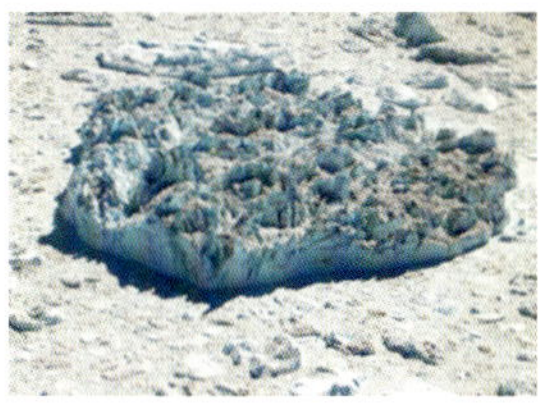
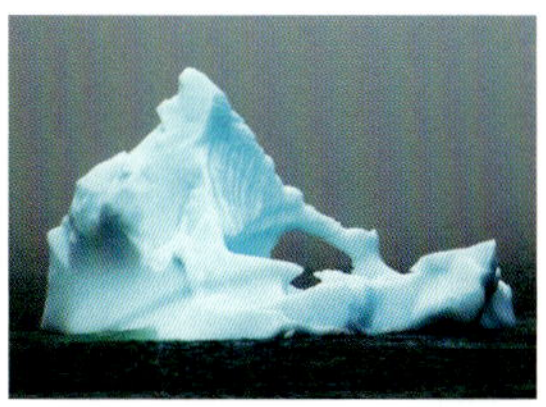
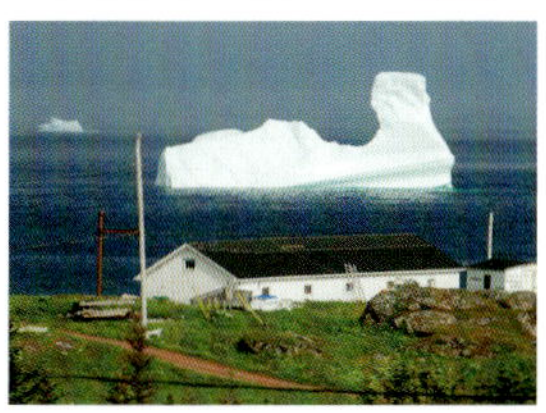
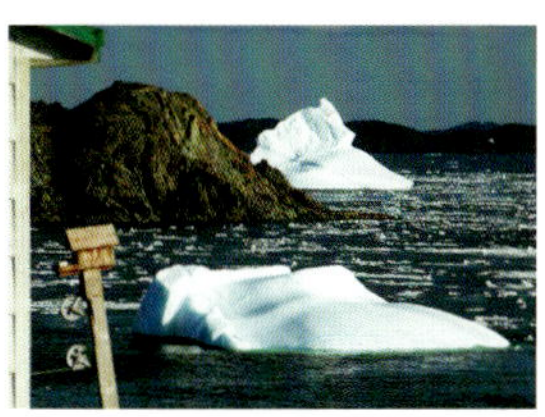
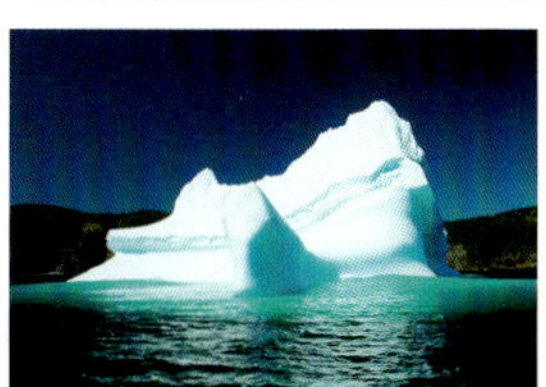

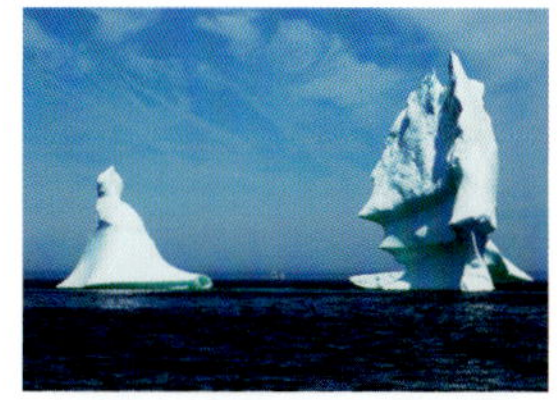
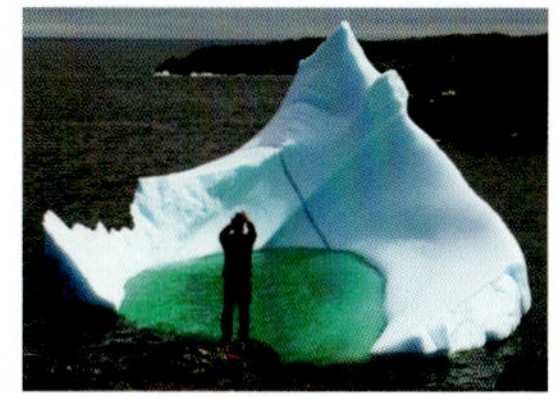
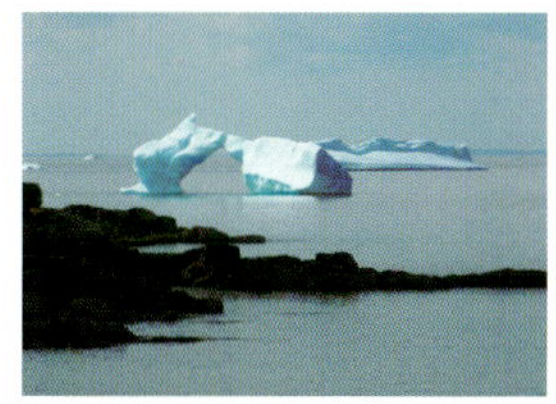
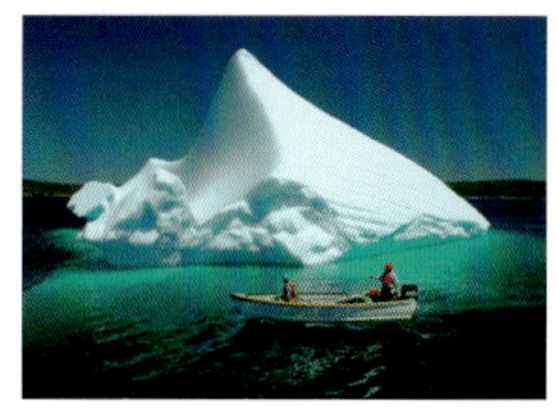
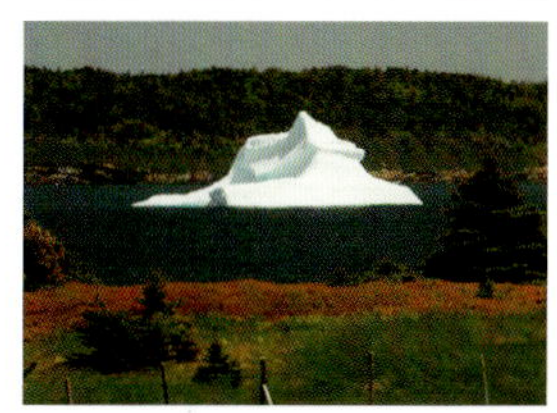
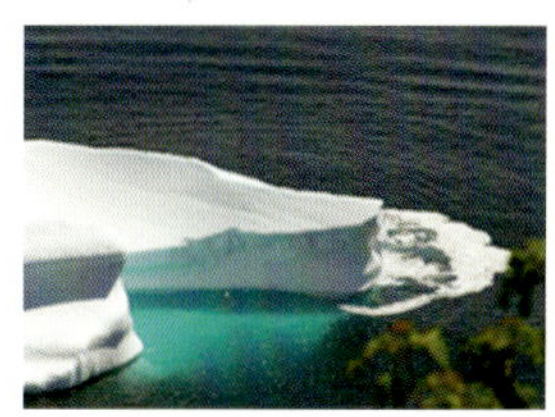
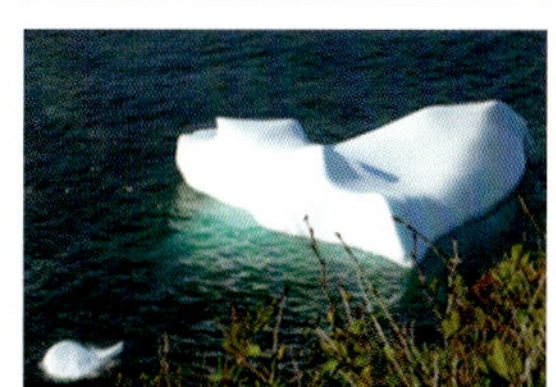

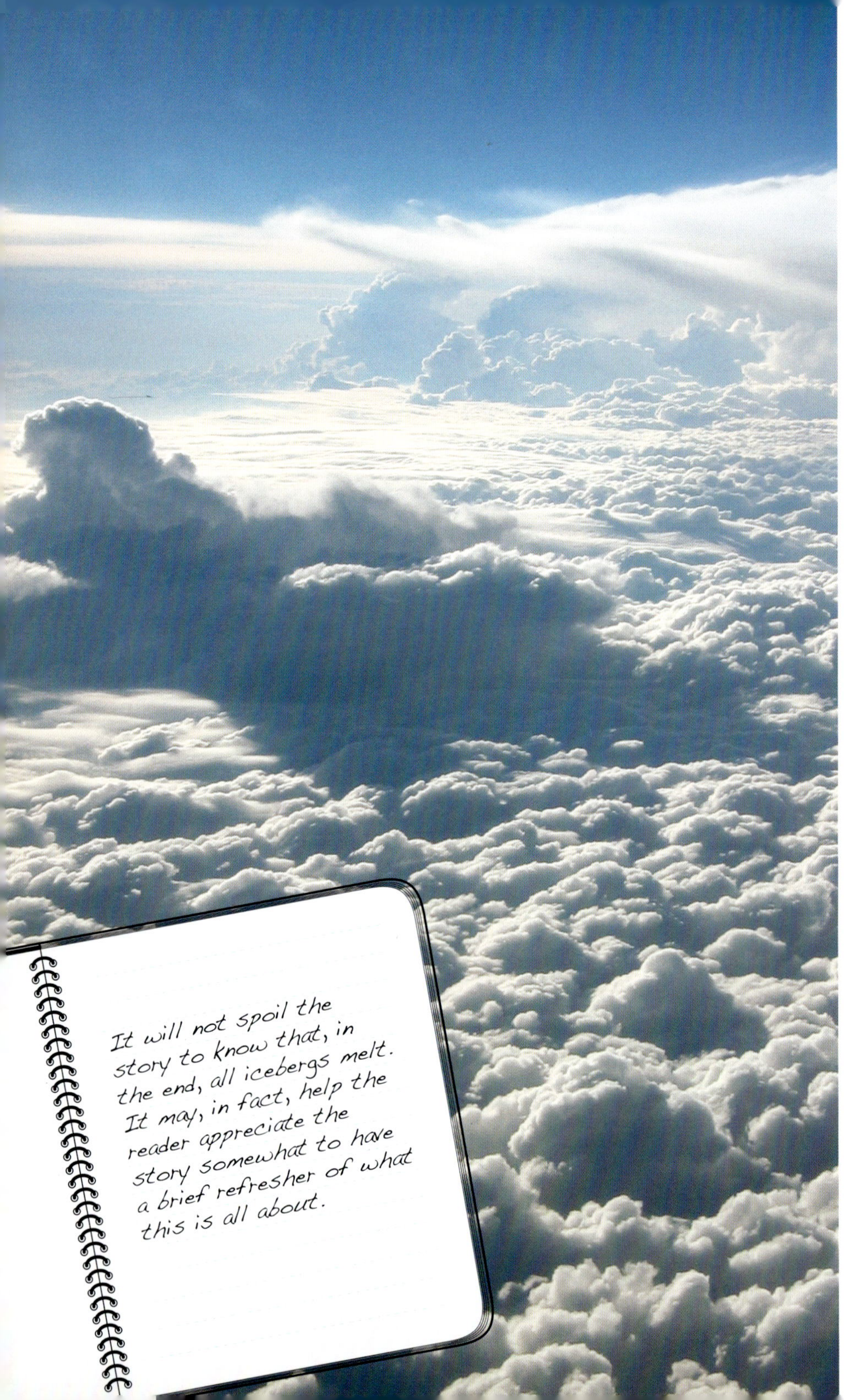

Introduction

Snow falls on distant arctic lands, over millennia becomes glacial ice that creeps to the coast and breaks off in mountainous chunks that float away – often taking so long to melt that a few survive to drift thousands of kilometres south into the domain of people who work and live oblivious to the faraway circumstances and improbabilities that give rise to this unique spectacle but are nonetheless awestruck by the natural beauty and excited by their fleeting presence. We call these floating things icebergs, and the path they take – Iceberg Alley.

The photos and experiences described in this book were for the most part collected over a five-year window by both authors working collectively. Our quests ranged from hour-long reconnaissance drives to week-long expeditions to Labrador and beyond. Word of a phenomenal iceberg here or there was usually enough to scramble one or the other of us into action, and often, paid work and family vacations were surreptitiously influenced by iceberg reports. The end result is this book which through a slight shuffling of chronology follows the course of icebergs from their very source in Greenland, to their most distant point of disintegration as "growlers" around the coast of Newfoundland.

Though not all arctic icebergs calve from Greenland glaciers, the vast majority do, and many of these come from a particularly productive fjord near Ilulissat highlighted herein. Not every berg makes it to Newfoundland either, but lots do and every single one ultimately reduces to the size of a growler before vanishing altogether. It is these things that we have seen, and so it is with that authority that we stake our claim to have witnessed Iceberg Alley - from glaciers to growlers.

It might amuse some to know that on average for each picture that appears in this book, hundreds, if not thousands, were taken. The excitement of a chance or even planned encounter is often palpable; the shutter becomes irresistible. You see, our subject is, for the most part, a unique colossus that is on the move, precariously hanging onto existence like a specatular sunset. Not everybody will get a chance to see an iceberg, and certainly no one will ever get a chance to see any of these icebergs again. For that is part of the attraction: the images in this book capture moments and vantage points that are now inaccessible. Though the images may alone convey the beauty and awesome scale of our subject, it was our hope that by providing personal vignettes as we have - that we may also open the door to the excitement and wonder that surrounded each encounter on our journey of discovery down Iceberg Alley.

Steve & Kevin

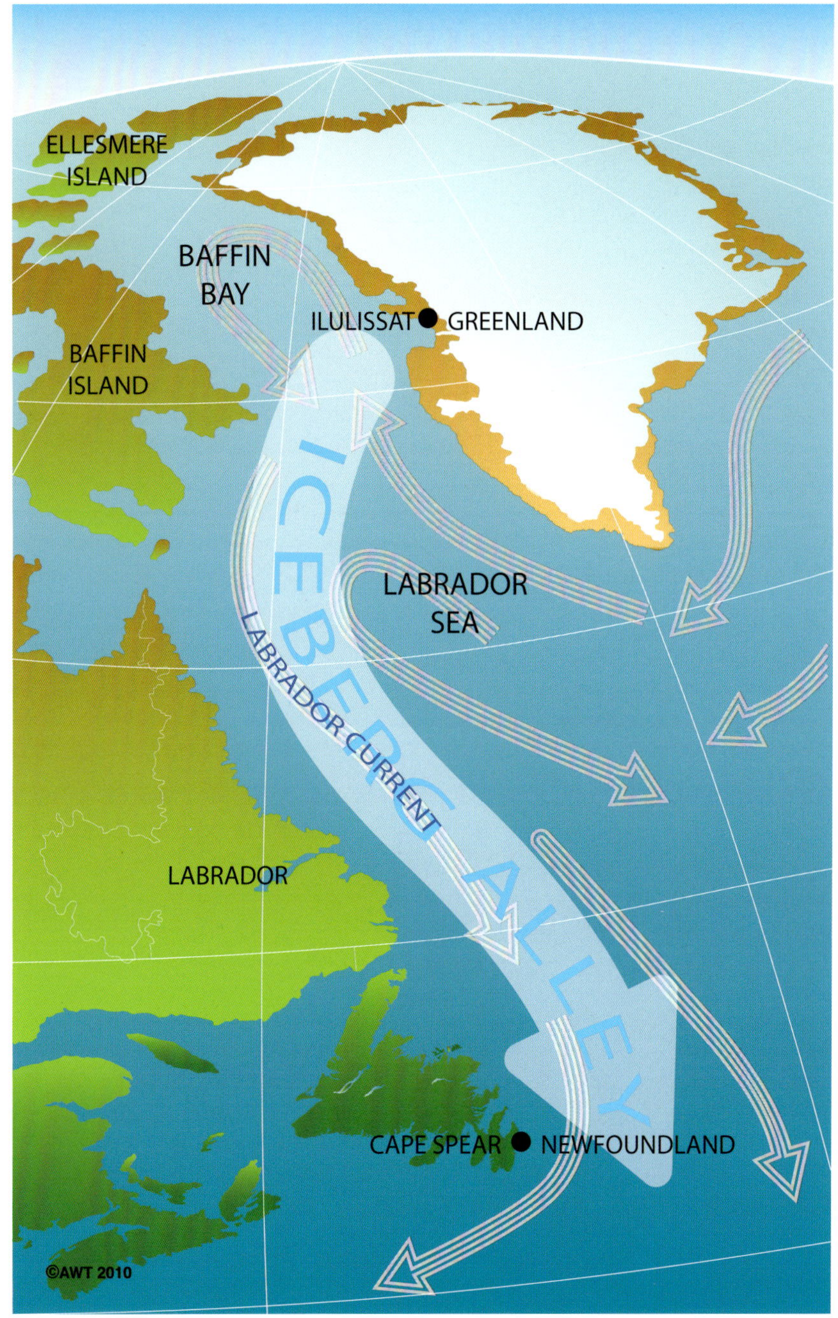

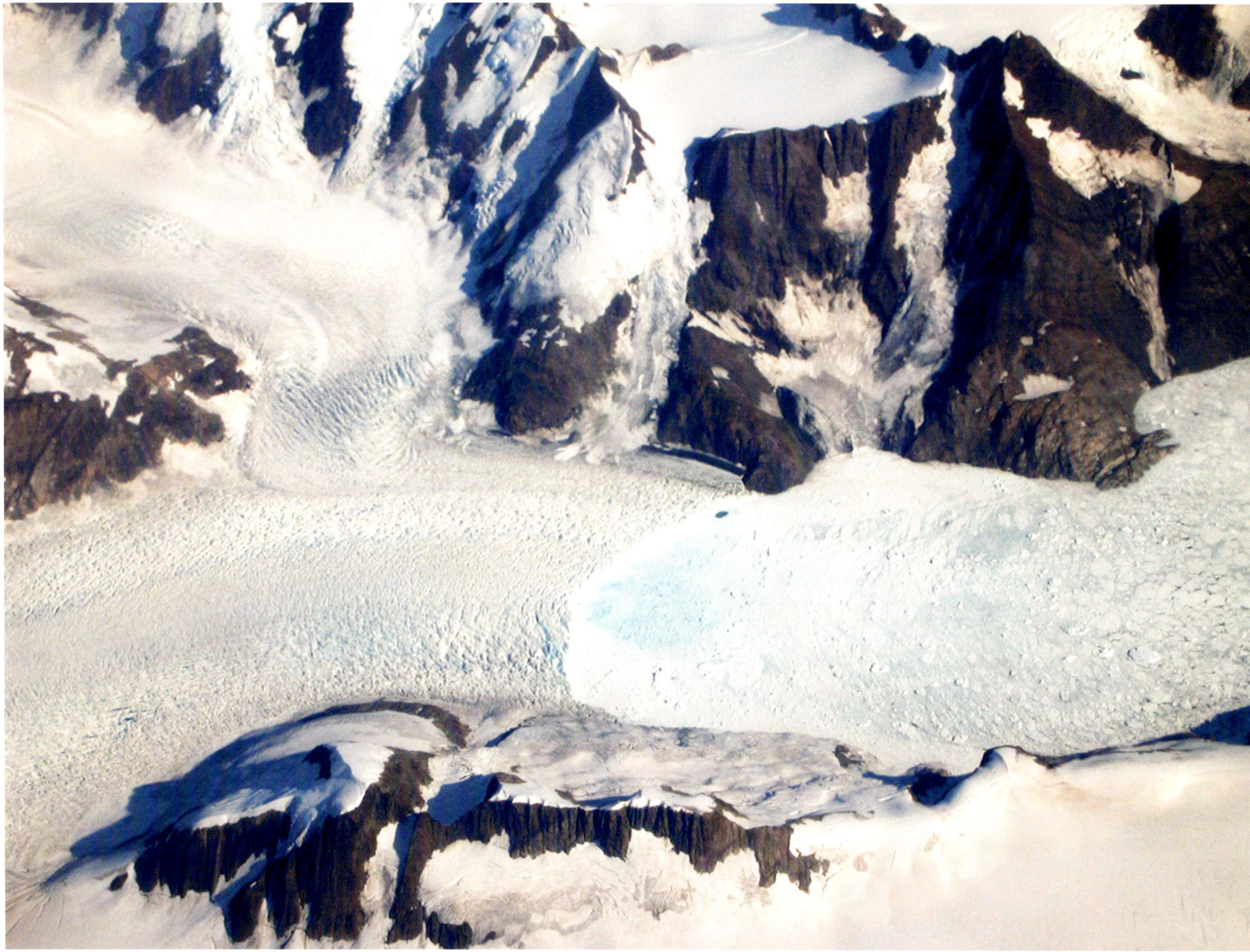

Greenland

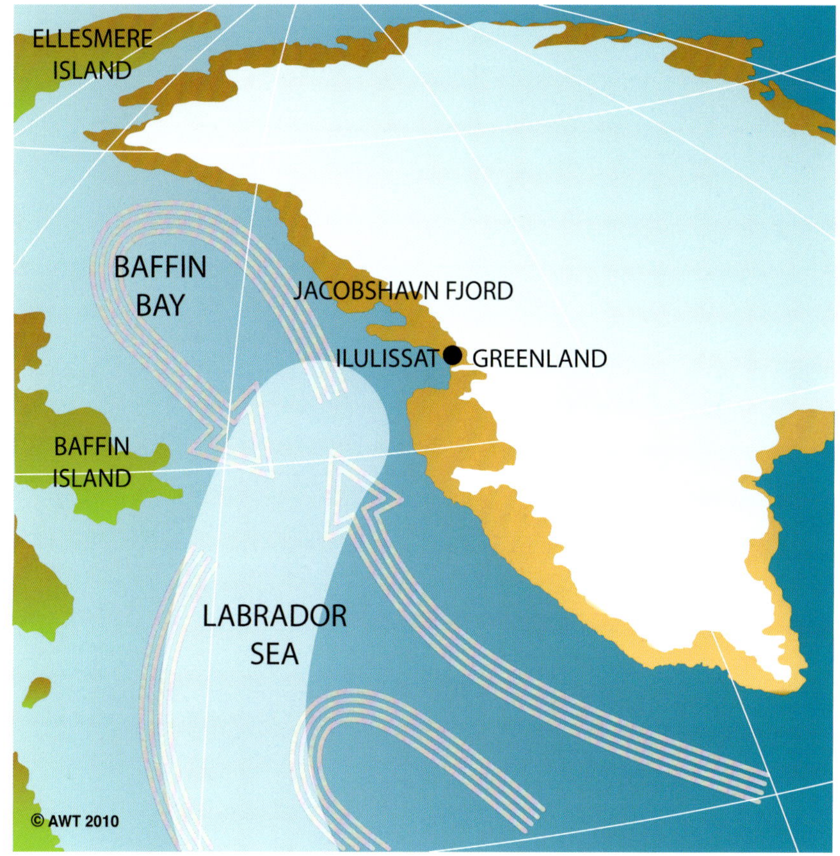
ELLESMERE ISLAND
BAFFIN BAY
JACOBSHAVN FJORD
ILULISSAT
GREENLAND
BAFFIN ISLAND
LABRADOR SEA
© AWT 2010

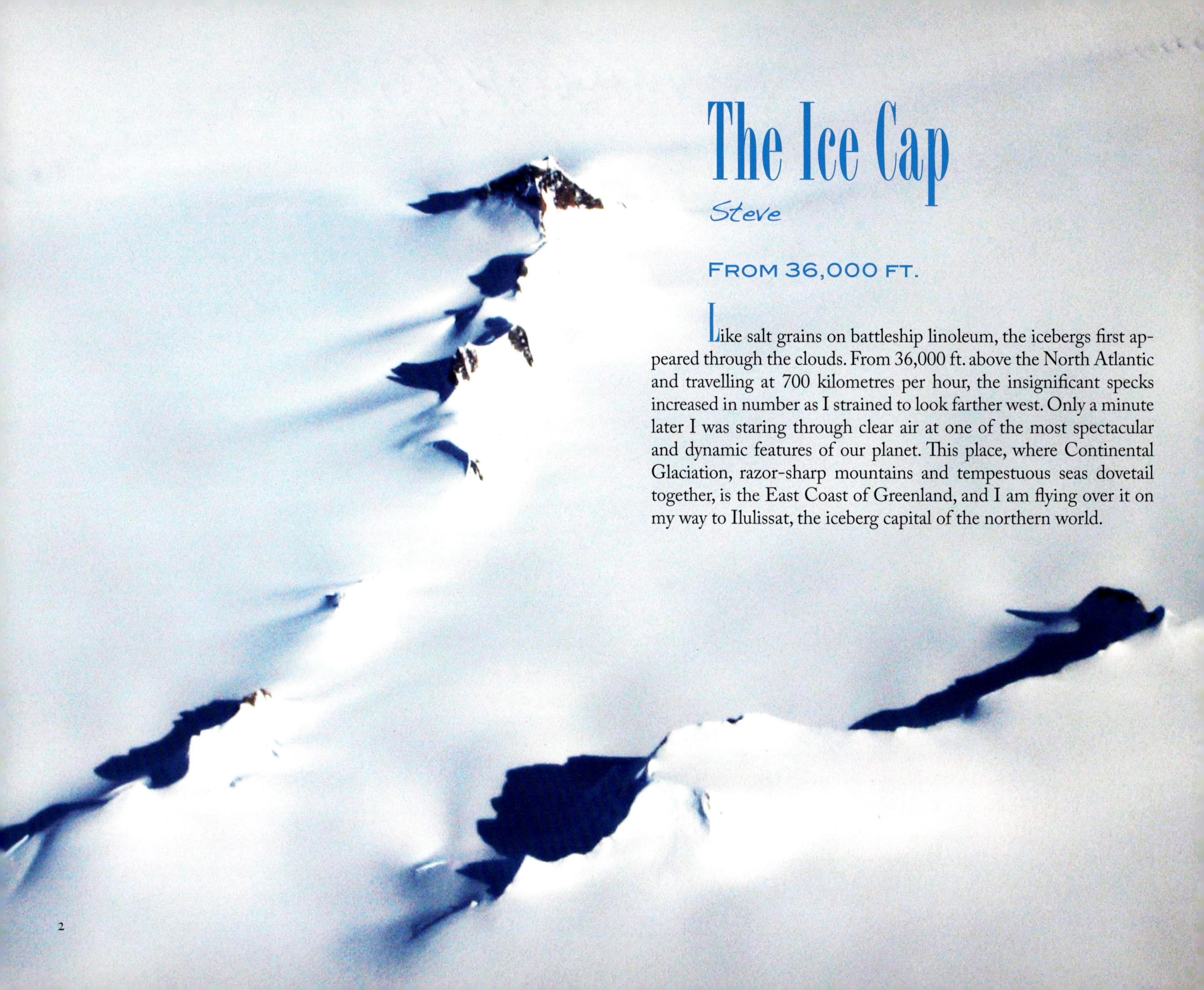

The Ice Cap

Steve

FROM 36,000 FT.

Like salt grains on battleship linoleum, the icebergs first appeared through the clouds. From 36,000 ft. above the North Atlantic and travelling at 700 kilometres per hour, the insignificant specks increased in number as I strained to look farther west. Only a minute later I was staring through clear air at one of the most spectacular and dynamic features of our planet. This place, where Continental Glaciation, razor-sharp mountains and tempestuous seas dovetail together, is the East Coast of Greenland, and I am flying over it on my way to Ilulissat, the iceberg capital of the northern world.

It was shocking to regard the magnificent mountains along the coast giving way to the relentless white as we flew inland. The mighty range shrank as if flooded by a sea of snow until only a few scattered black tips protruded anomalously.

I chuckled at the cartoon that appeared in my head where a polar bear striding past one of these peaks turns to another and says "they say 7/8 of it is below the surface . . . ."

Soon there is nothing but white. No longer possible to distinguish between snow and clouds, the surface just rounded over and its vastness was revealed.

The first sign of change was an odd blue strip that first appeared like a stain on the white quilt below. But very quickly it became apparent that there were others and that they were connected. These lakes of meltwater set on top of the vast Greenland ice cap were so aquamarine that they appeared to be a cheap fake; some still had thin surface ice drifting about pointlessly.

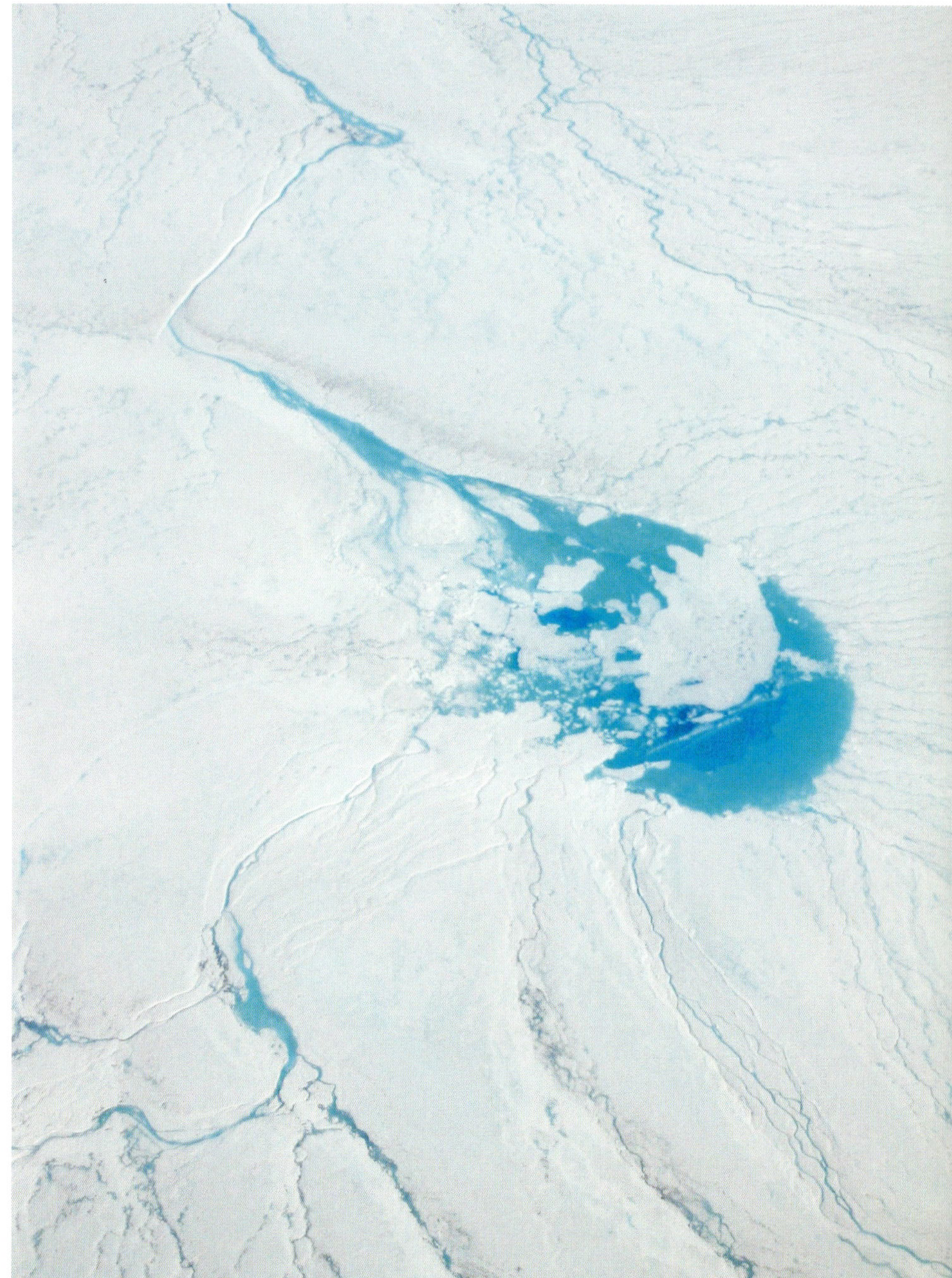

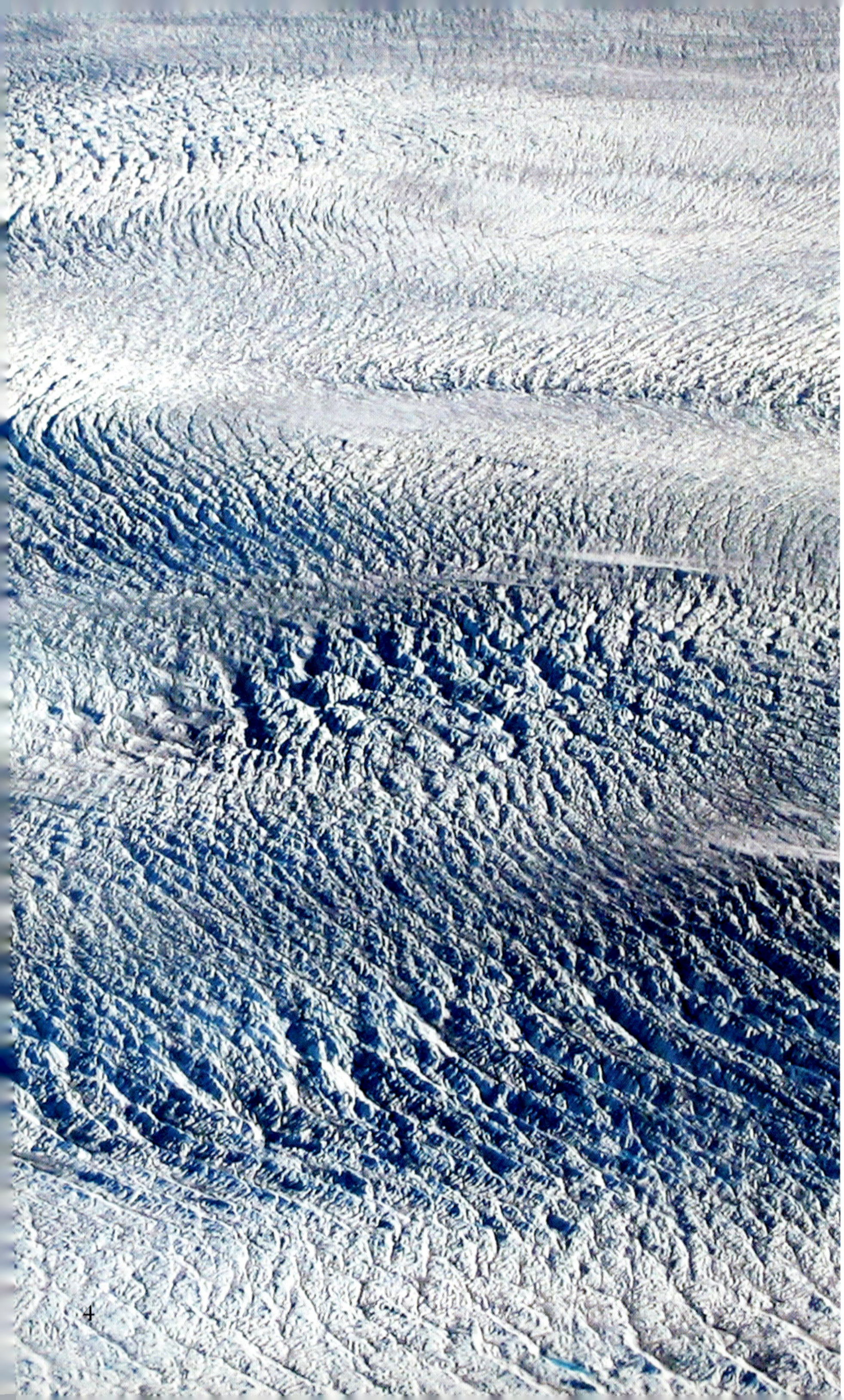

A Glacial Place

Steve

The next scene was more desert-like than anything, a scoured terrestrial surface with wrinkles and cracks, dry riverbeds and canyons like an endless icy badland. As the plane flew on toward the lower western edge of the ice cap, the surface took on a grim look, the contours were deeply fissured, lines were twisted and the surface despicably dirty.

The rough features narrowed into flow-like patterns, and straining to look ahead I could see why. Land ahead on the leading edge was blocking the apparent creep of the icy mass, forcing it into river-like channels.

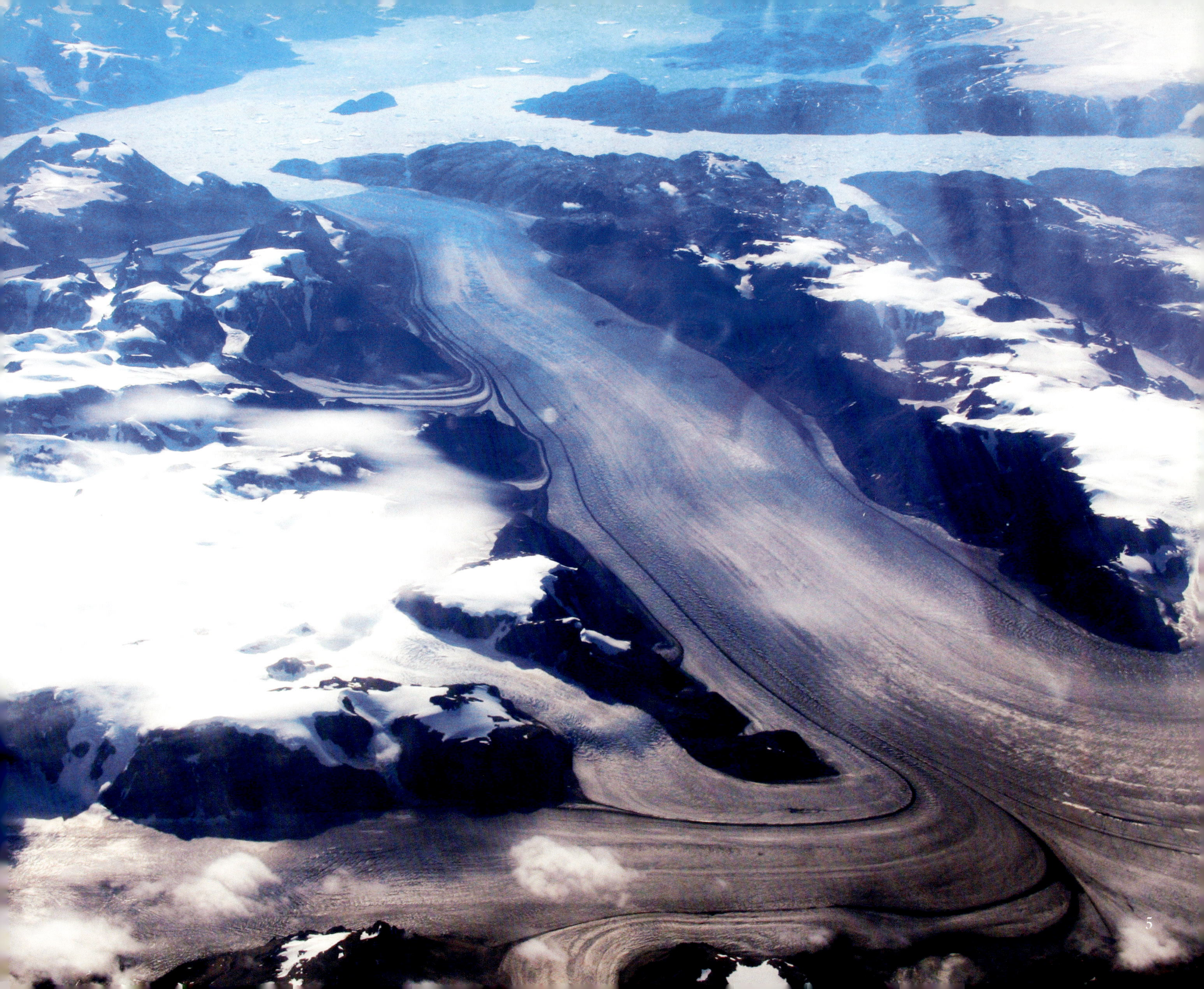

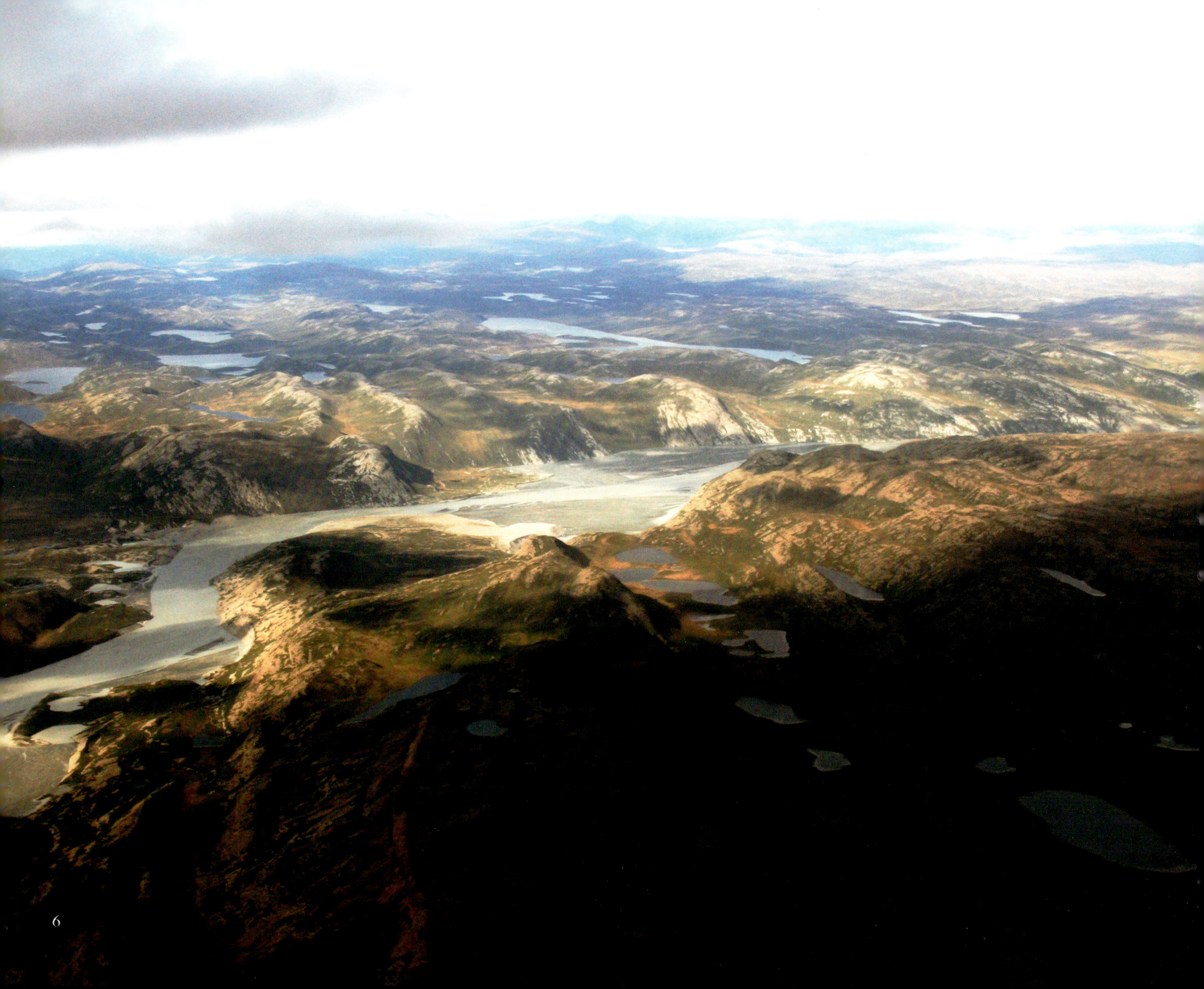

The Green Land

Steve

Though Captain Bob Bartlett was right when he once quipped that whoever named Greenland must have been in the real estate business . . . I was quite surprised at just how green parts of it were when flying north over the western shore of that Great Island.

It was at that time that I witnessed what looked like an iceberg conversation. Not a conversation about icebergs, a conversation between icebergs . . . . There were only three or four icebergs in the fjord below, positioned like chess pieces a few squares apart on a jade-green playing board. Radiating outward from one of the bergs were very smooth concentric circles, looking like sound waves drawn by a cartoonist. The curved lines that intercepted the other bergs were likewise reflected in opposing directions, giving the impression of a deliberate response in an otherwise dead-silent setting. Well, the image passed by fairly quickly and I soon surmised that there had been an instability in the main iceberg causing it to rock gently without any violent calving or rollover.

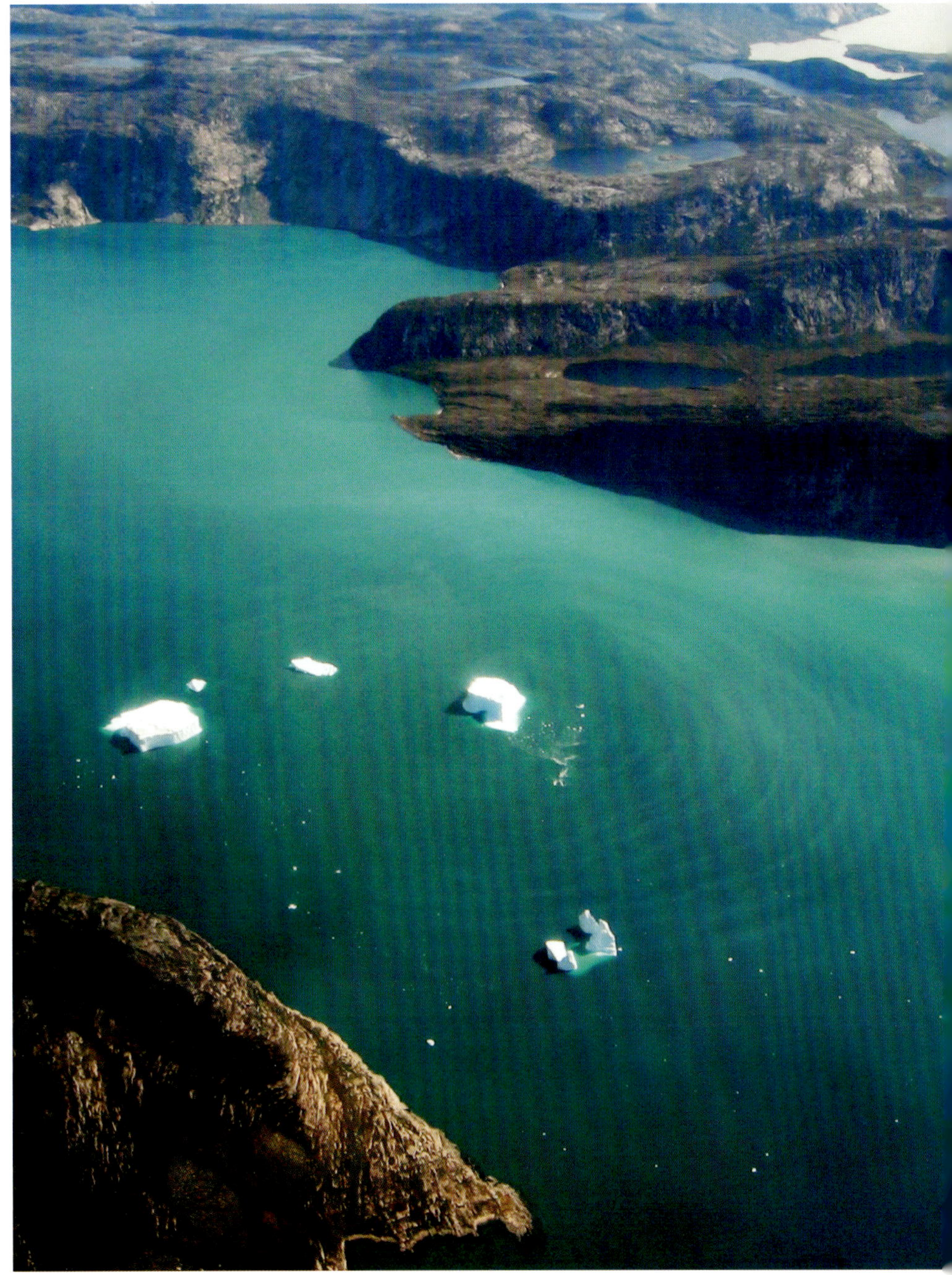

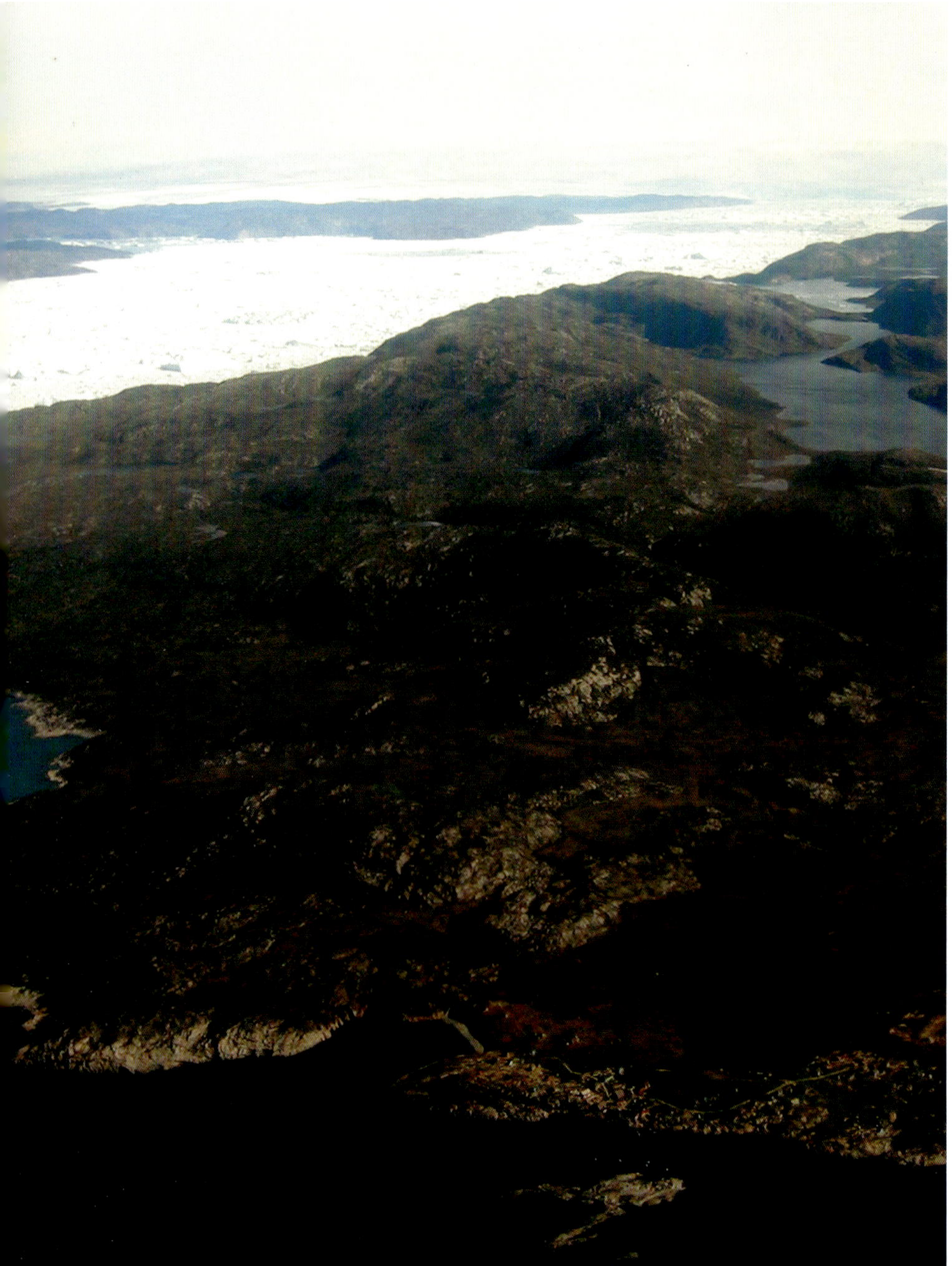

Icebergs Galore

Steve

Continuing north aboard Air Greenland, my feelings turned to worry as we passed fjord after fjord of emerald green water and rugged mountain glade with only a few bergs in sight. I fumbled with the map and cringed at the thought that I had somehow misinterpreted the iceberg literature about the amazingly productive glaciers in this region. Then as I strained to look forward through the oval glass I saw a distant white ribbon that seemed to stretch from the ice cap to open water ahead. As the aircraft approached all doubt vanished – the jaw-dropping spectacle was unlike any landscape I had seen before or since. Before my eyes was an iceberg factory of tectonic scale: the vast white ice sheet on the distant horizon pouring into the jagged glacial front reminiscent of a frozen Horseshoe Falls at Niagara. This front was the starting line of a fifty-kilometre-long, ten-kilometre-wide conveyor completely packed with icebergs of all sizes. The stream of bergs loosened as the fjord gave way to a massive delta of colossal icebergs apparently too deep to pass over the fjord sill and thereby temporarily prevented from entering the deep blue waters of Disko Bay. Had the plane turned homeward-bound at that point, I would not have left disappointed.

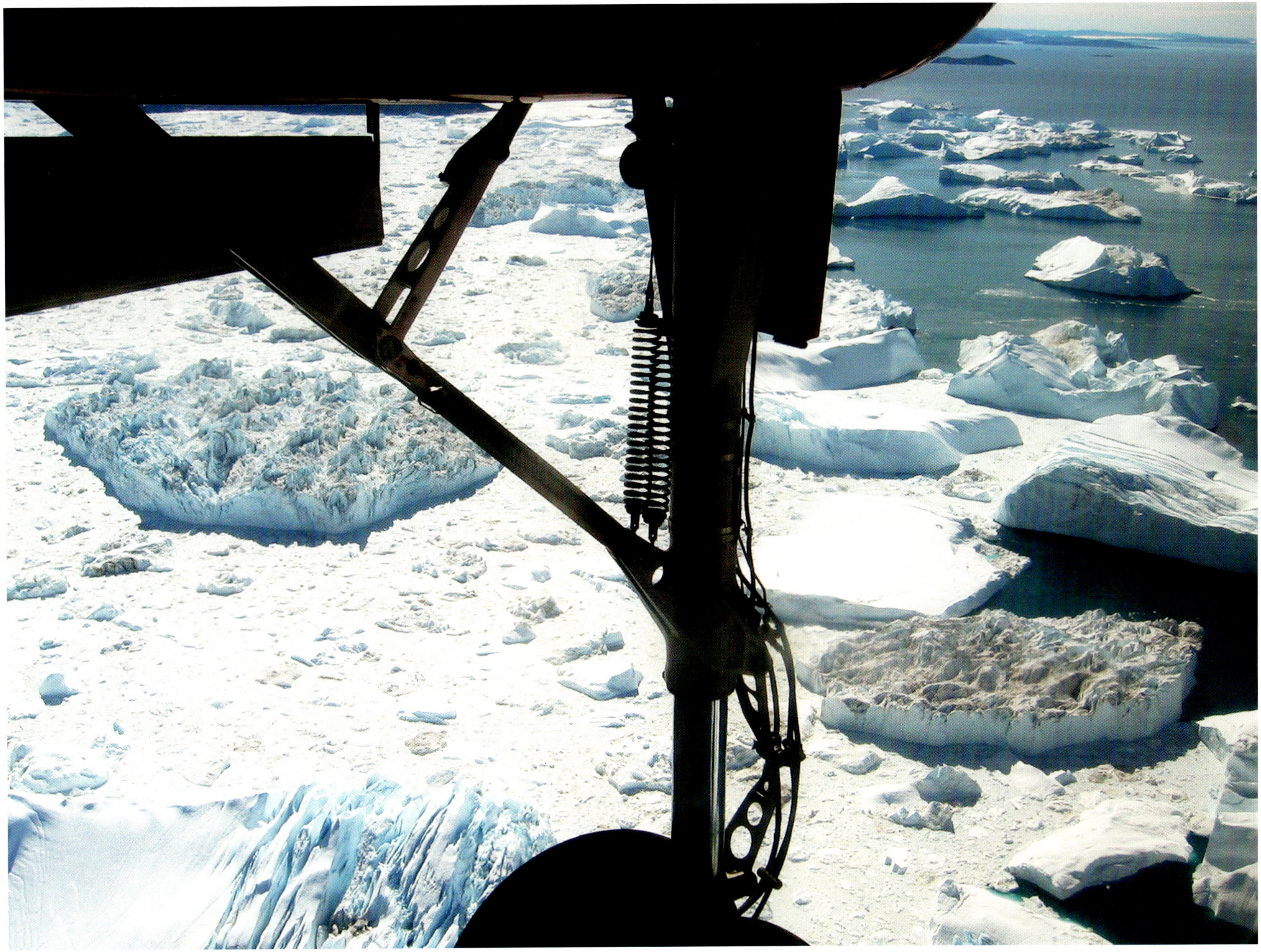

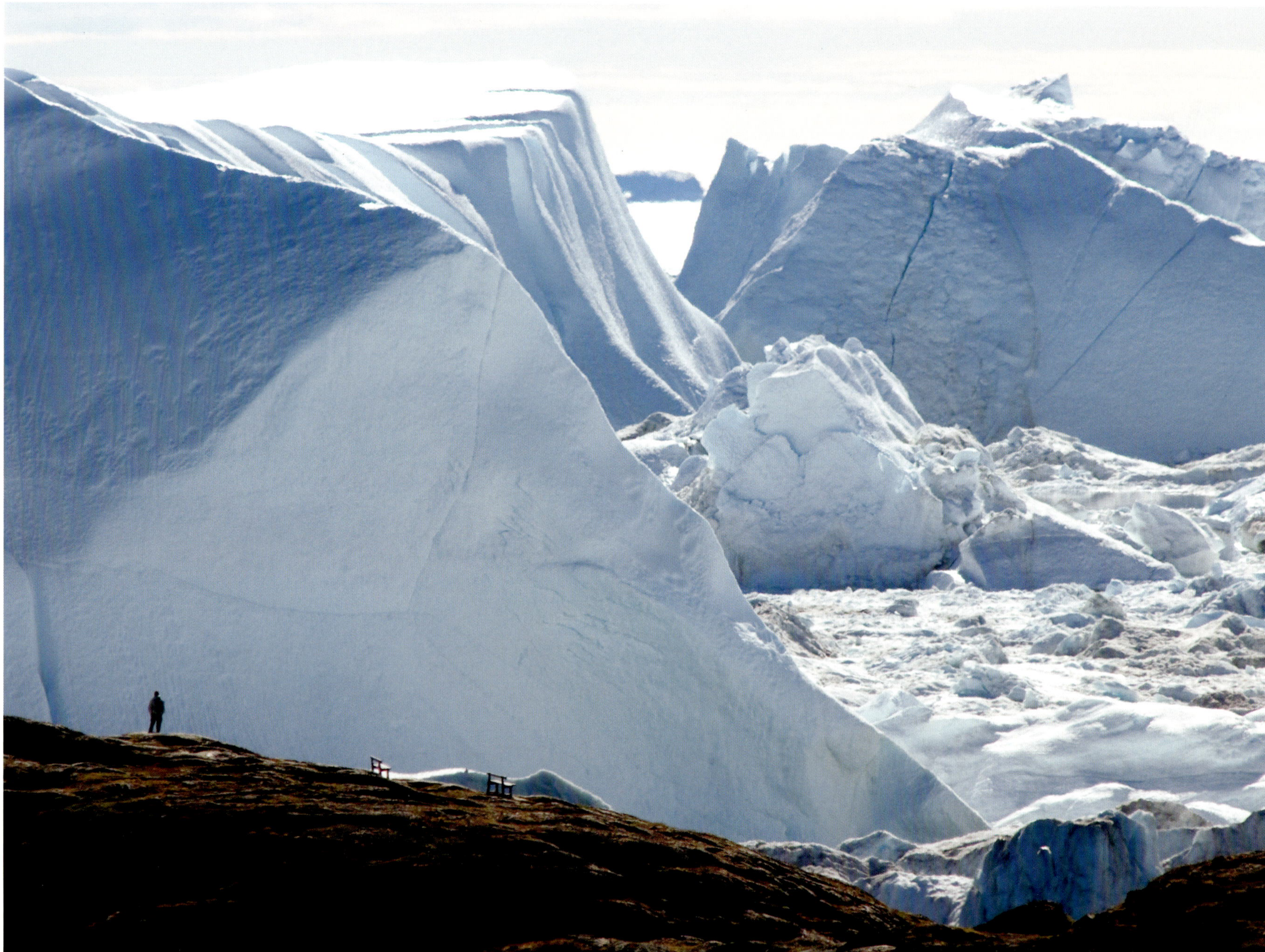

The Pole Position

Steve

The town of Ilulissat (meaning "iceberg" in Greenlandic) sits around the corner from the mouth of the fjord alternately referred to as the Jacobshavn (Danish) or Ilulissat Icefjord. The outlet glacier where icebergs are calved is positioned fifty kilometres east of the town and is called the Jakobshavn Isbræ or Sermeq Kujalleq in Greenlandic. Remarkably it is from this single location that approximately thirty-five cubic kilometres of ice accounting for ten percent of all icebergs produced annually in the Northern Hemisphere originate."

The only feasible means for tourists to get a close look at the calving glacial front is by helicopter, an expensive inconvenience for tourists who yearn to get to the focal point of the iceberg phenomenon. This extravagance proved irresistible to my travel companion (my father, Angus) and I did not reject his invitation to accompany him.

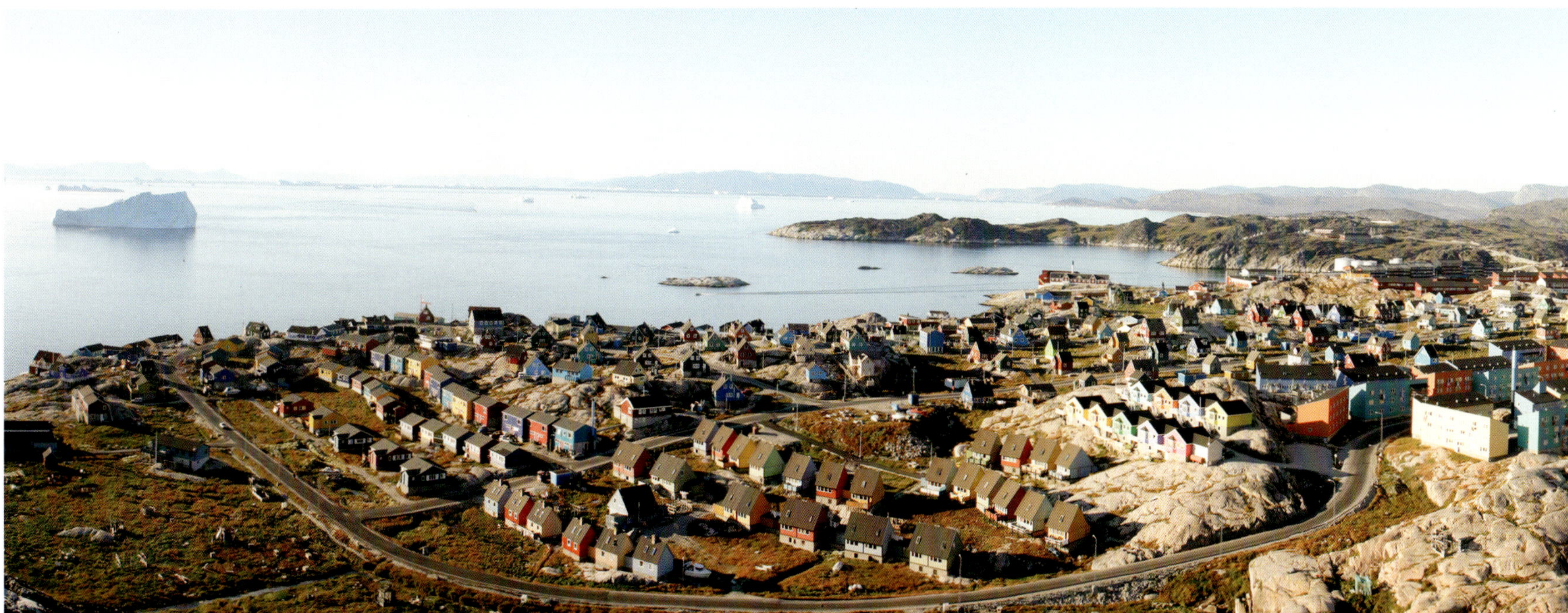

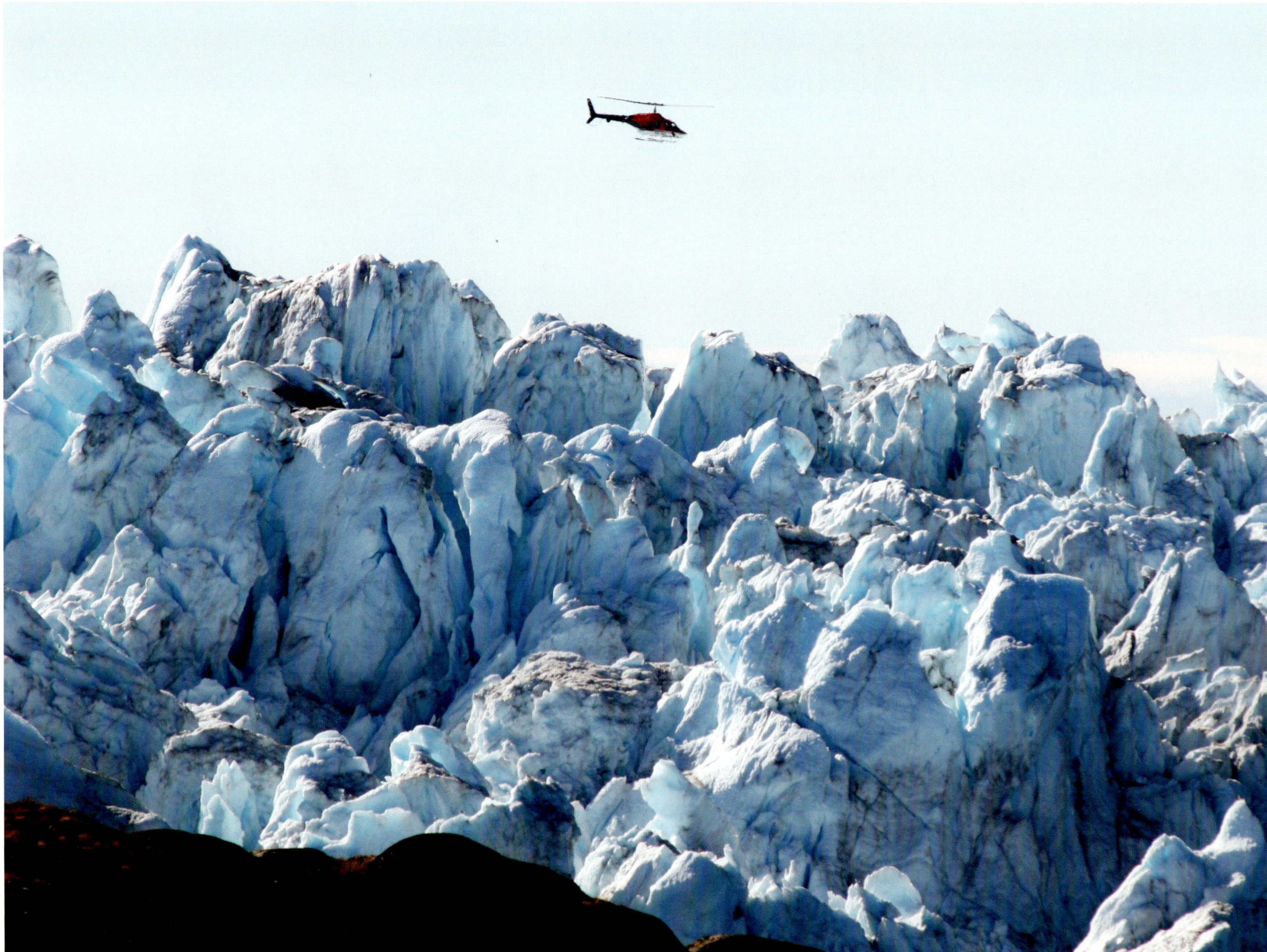

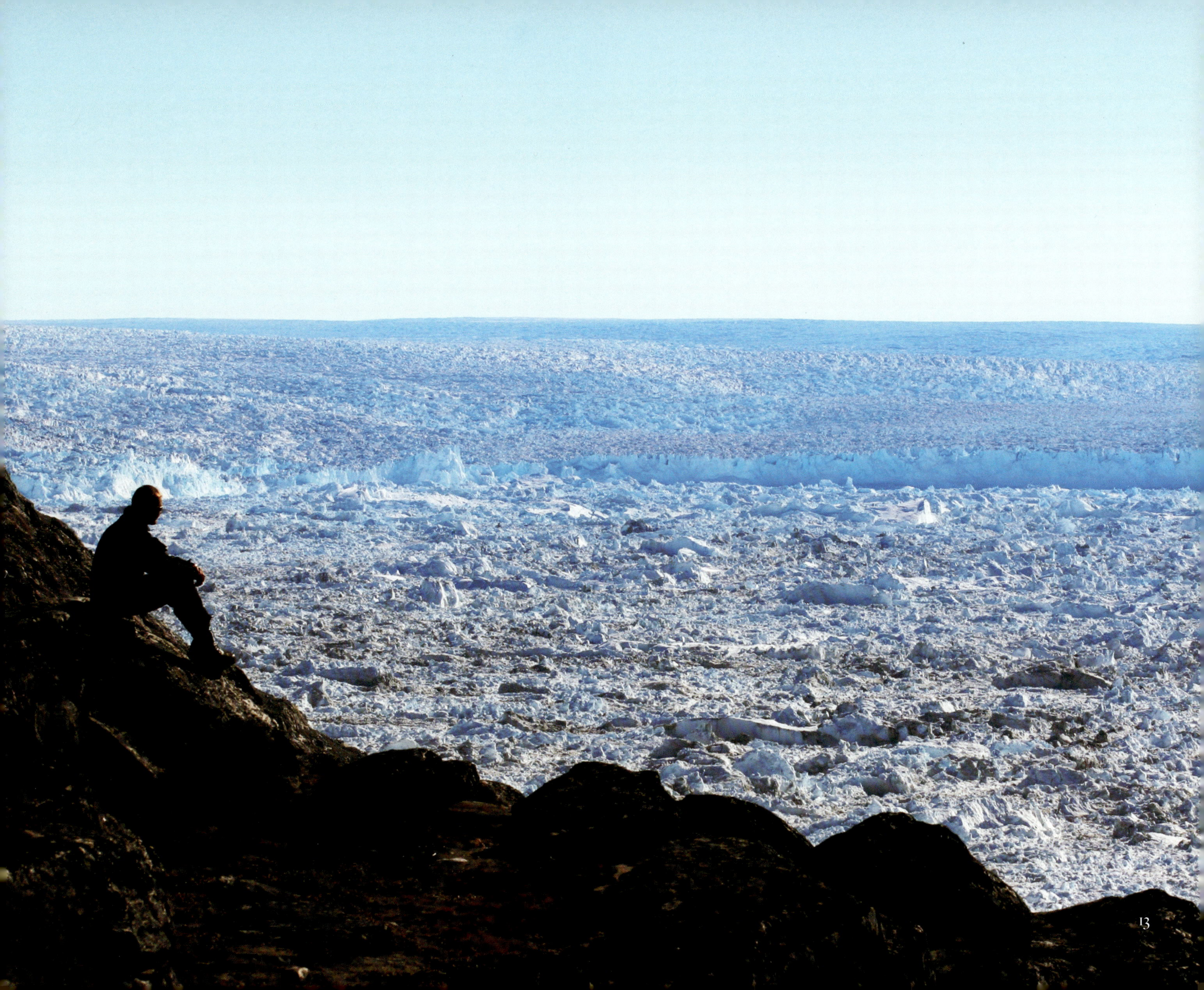

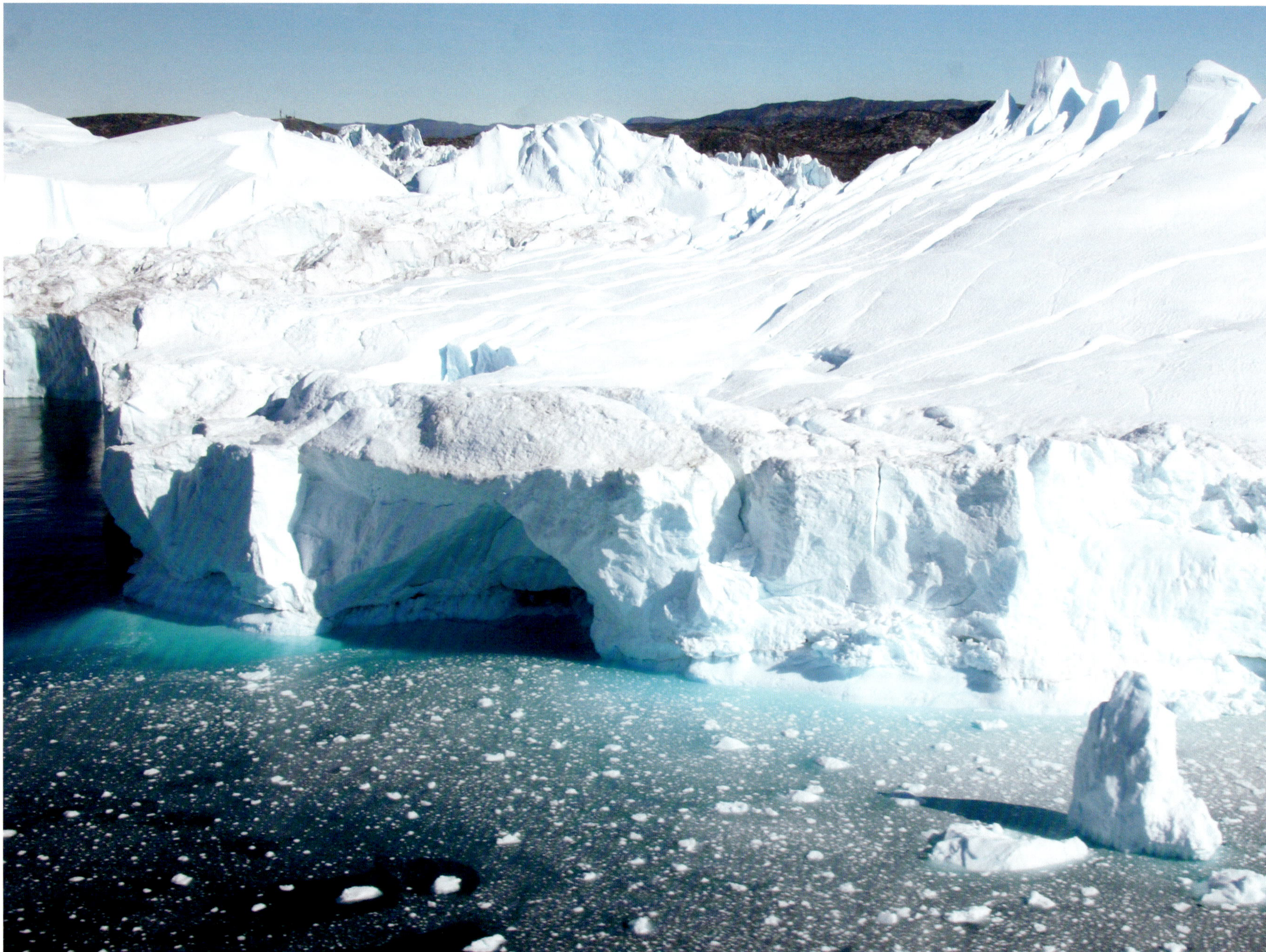

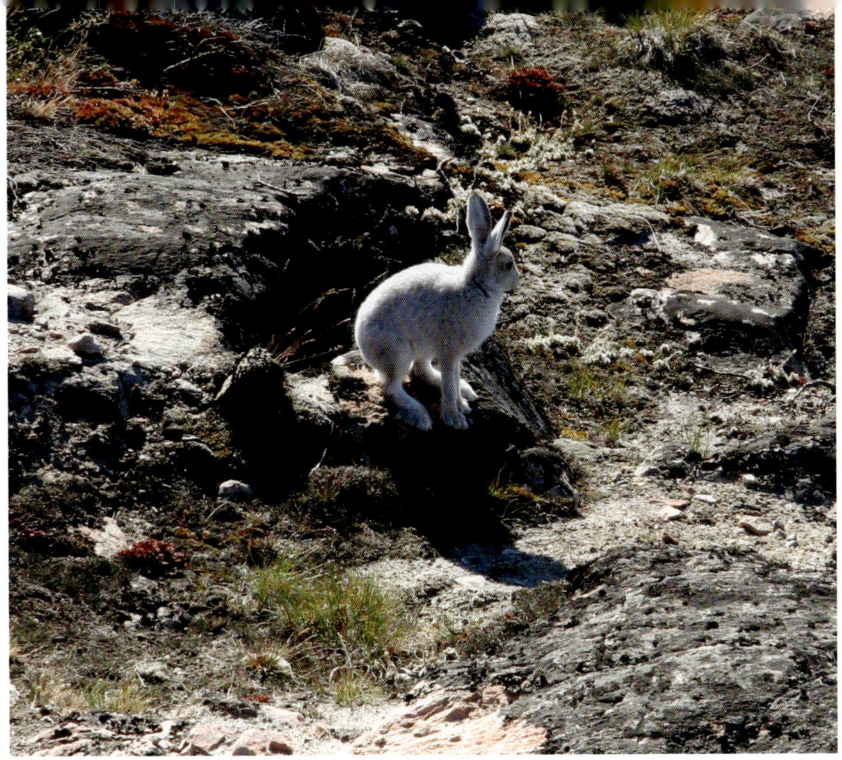

The ride did not disappoint – there was a spellbinding array of giant glacial and geological features, framed by a brilliant vast sky, and speckled navy blue water. Paradoxically, once on the ground nearest the glacial wall where the floating behemoths are brought to life, the air was silent, tiny wildflowers peeked through the strewn till, and a delicate arctic hare hopped by.

The return flight was spectacular as the helicopter pilot soared down the fjord plunging between giant bergs like a dragonfly. If not for the shadow of our craft, the scale of the whole thing would have seemed at that point impossible to register. In half an hour we wove the length of the fjord to open water – a distance that takes the ice a year or more to transit. As we passed between the giants grounded near the town at the exit of the fjord we remarked how different some appeared from the air as from the ground earlier that day.

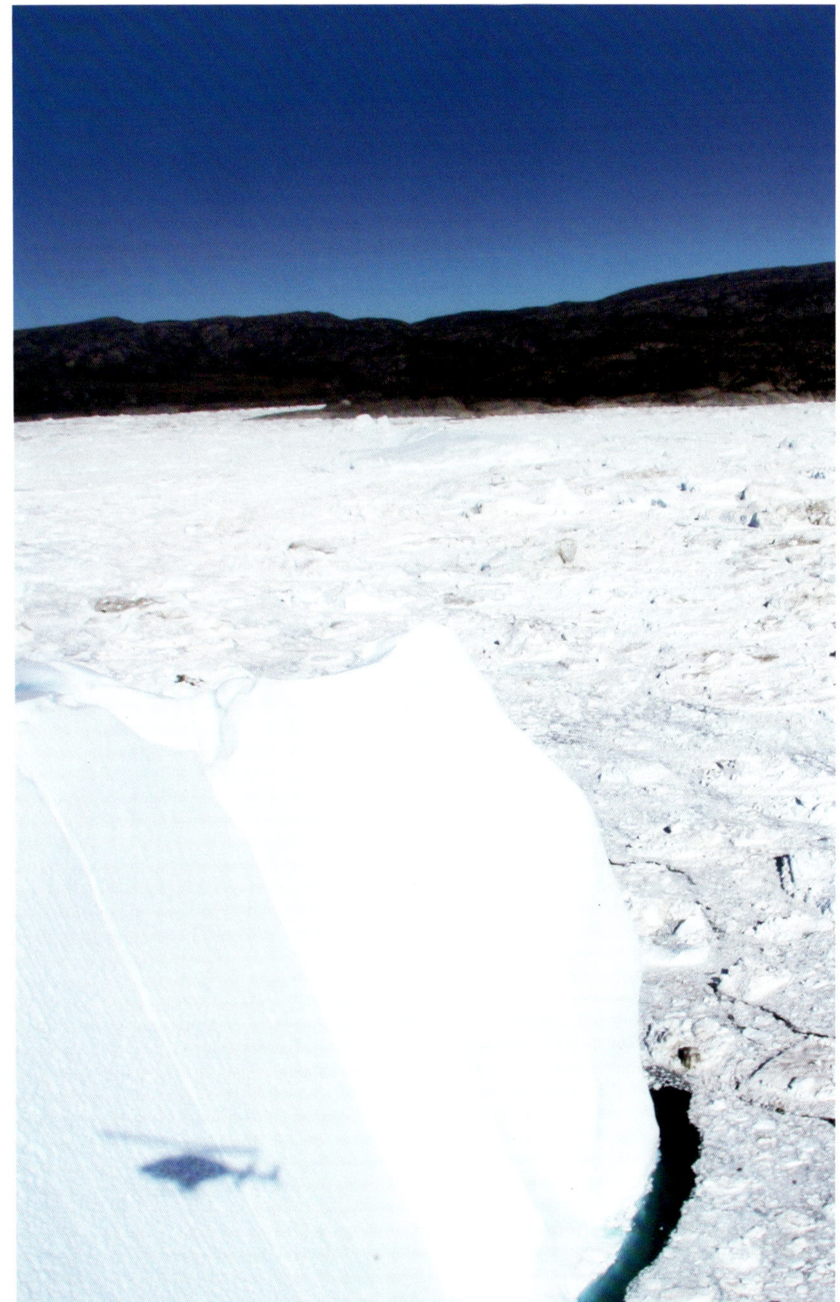

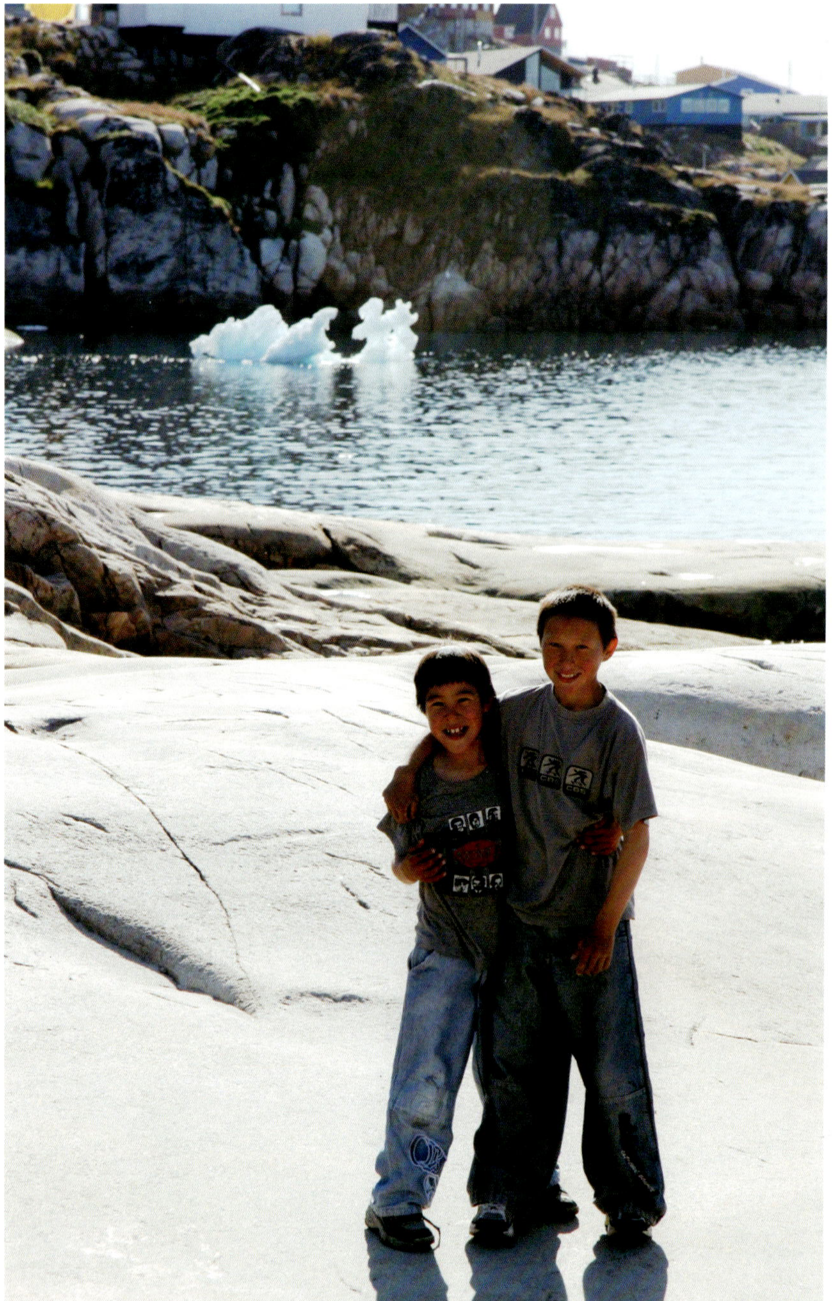

Life in Ilulissat

Steve

The seals thrive off the abundance of marine life around the Jacobshavn Icefjord. Apparently, the cold high-oxygen waters and nutrient-rich sediment continuously stirred up by the endless grind of icebergs yield high numbers of arctic cod, lance, and many other fish species. In turn, people have been living off the seals for a very long time. The ancient bone beds exposed at the beach near the town of Ilulissat give testimony to the means of survival of people living there in ancient times. I witnessed what must be an event no less common than a trip to the grocery store – a local fisherman cleaning a seal on the smooth surface of a bare granite rock on the shore. The crimson-red water around the rock was so brilliant that it would be considered unrealistic if it were a special effect in a movie. I noted a fairly large fish hook still dangling from the mouth of the flensed seal and received confirmation, using a universal hand gesture for a fish hook in the corner of my mouth, that he had indeed reeled in his catch by surprise while fishing for deepwater fish.

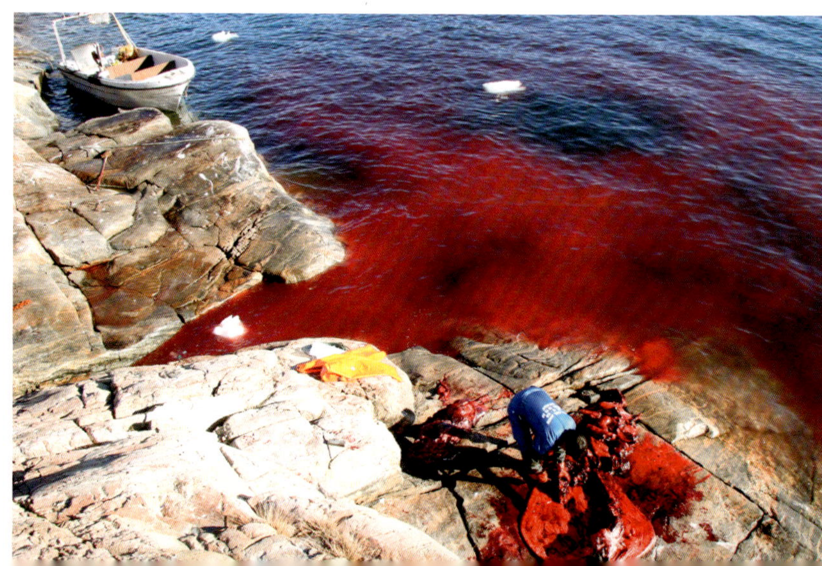

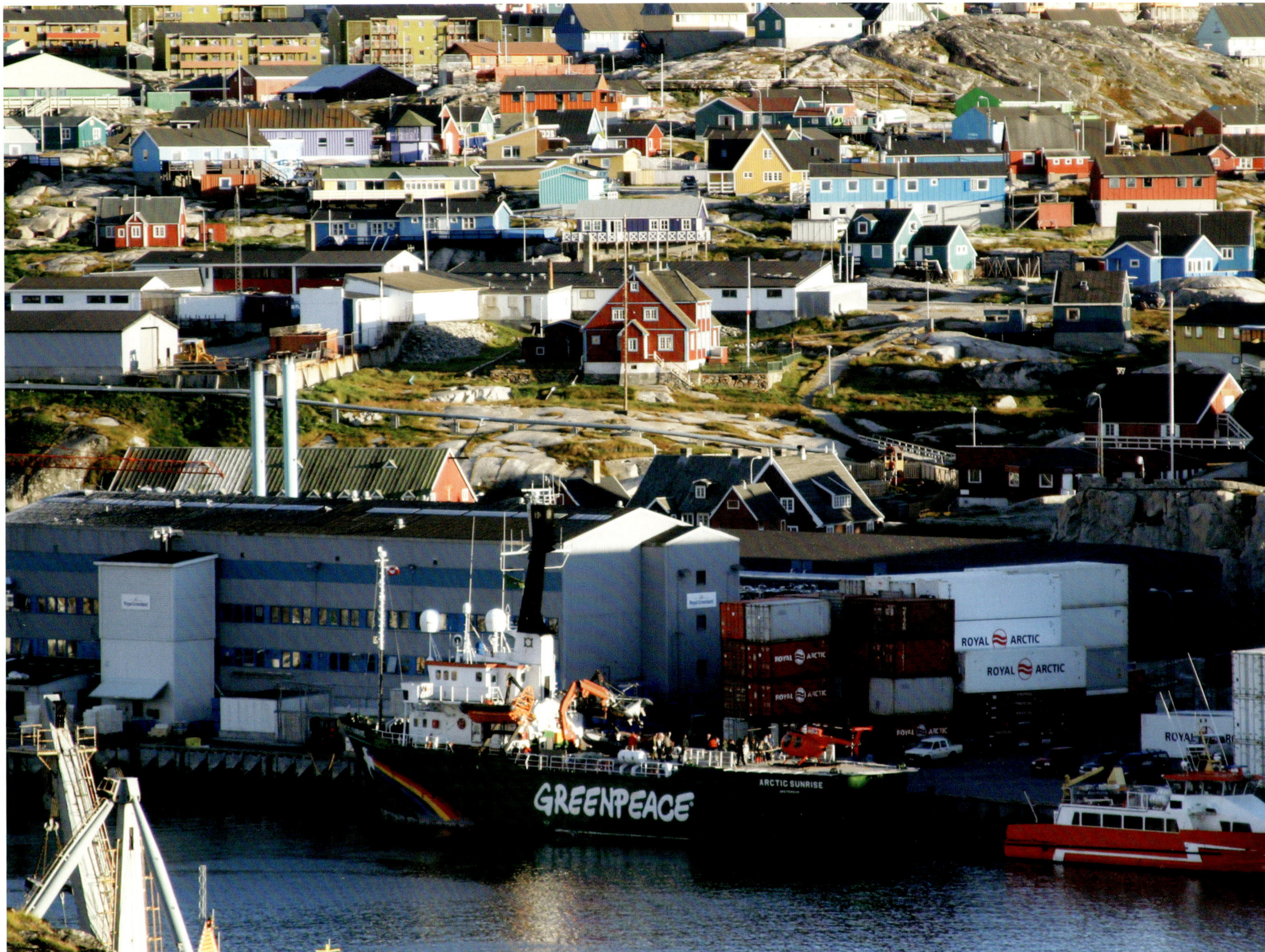
GREENPEACE
ARCTIC SUNRISE
ROYAL ARCTIC
ROYAL ARCTIC

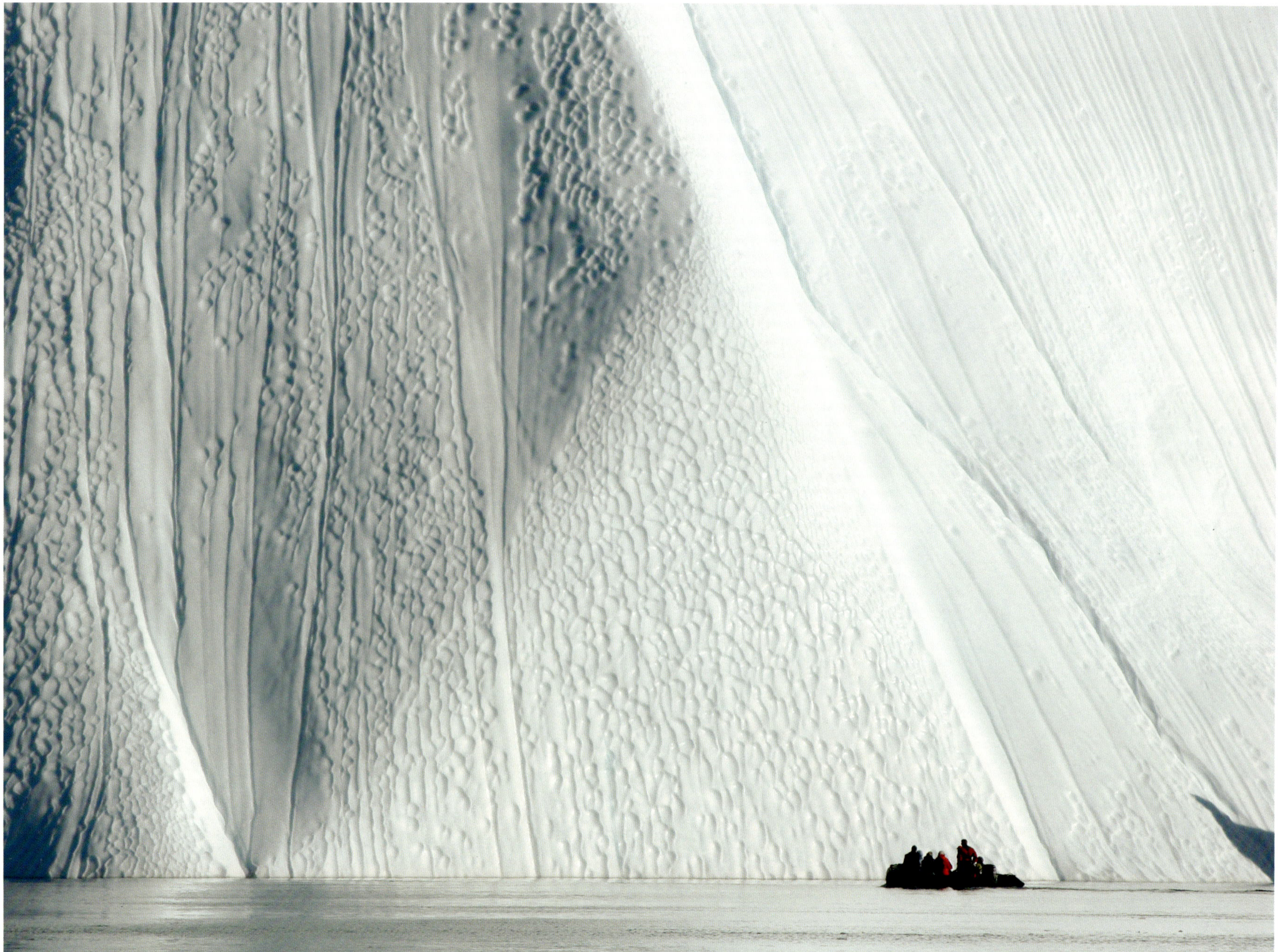

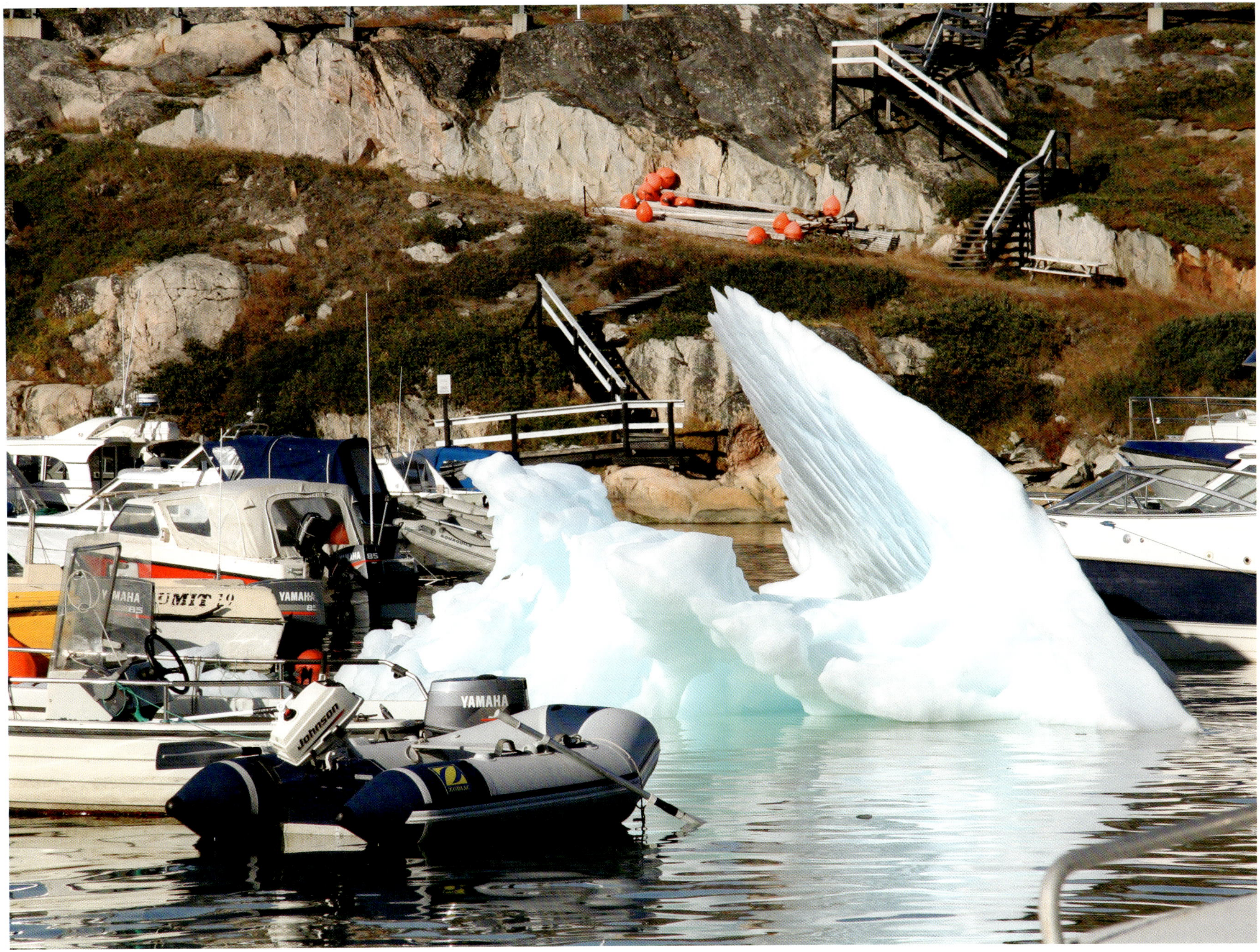
YAMAHA
YAMAHA
YAMAHA
Johnson

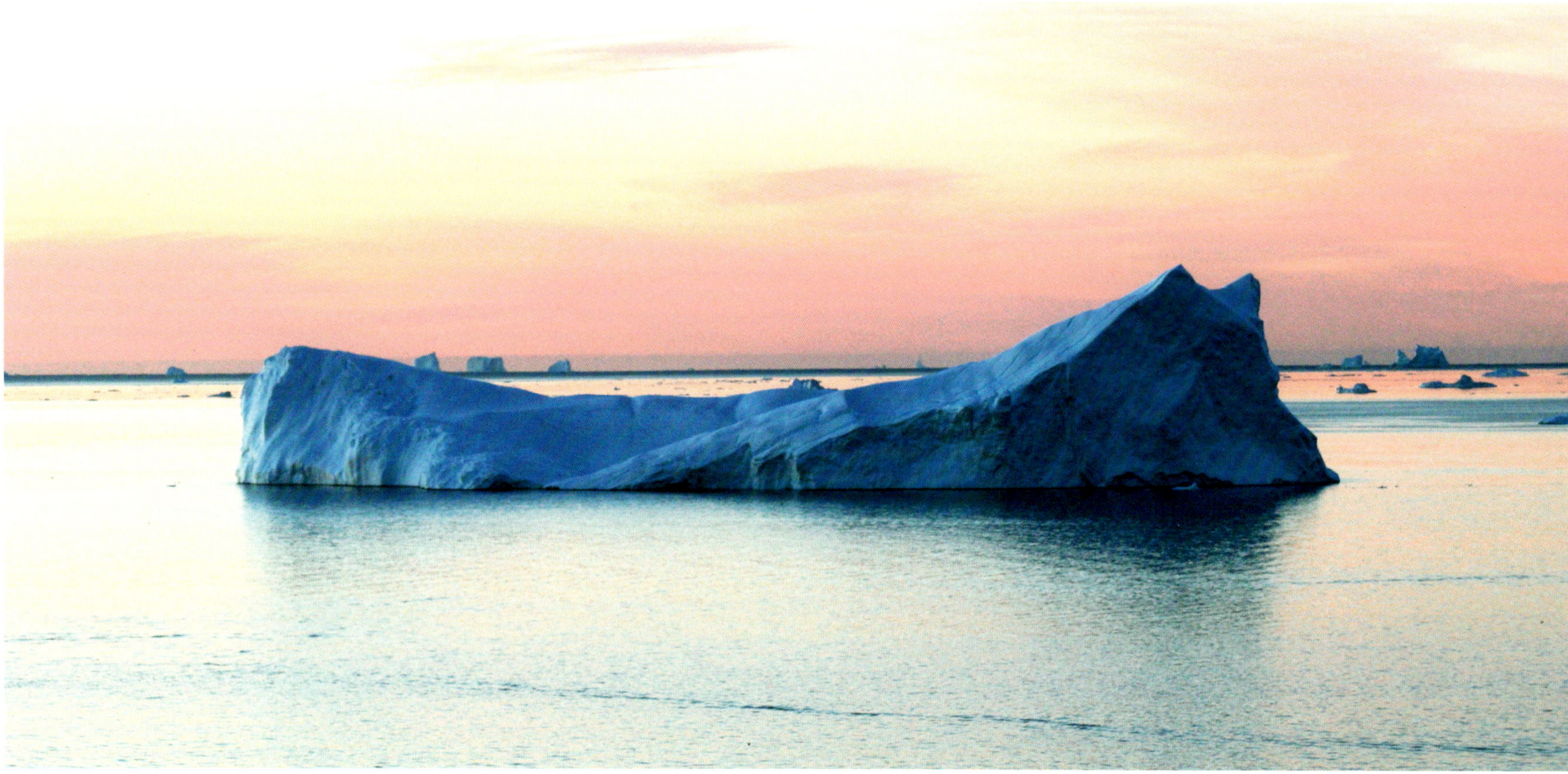

The Husky Berg

Steve

The berg we called Husky looked very much like the dog's head, or at least the dogs we thought were huskies. For the most part Greenland sled dogs don't look like the huskies I imagined they'd be. Apparently, these sled dogs, already unhappy and unclean due to the disappearance of snow, are even more miserable with lack of work and attention. Seems they are better kept this way in the summer, else they don't perform when duty calls in the winter. One thing they do seem to enjoy in the off-season is howling, though we weren't particularly bothered by this as the animals were mostly out of town while we were there.

The husky berg emerged from the fjord having spent its entire buoyant life of one or two years confined to the narrow rocky channel. In day one of its free existence it seemed to hang around the fjord mouth until, when a strong breeze arose, it pirouetted once, shot directly off shore, and then turned to the north. It seemed to delay its departure once again, and then without any apparent reason it sped off into the distance, making for the outlet of Disko Bay.

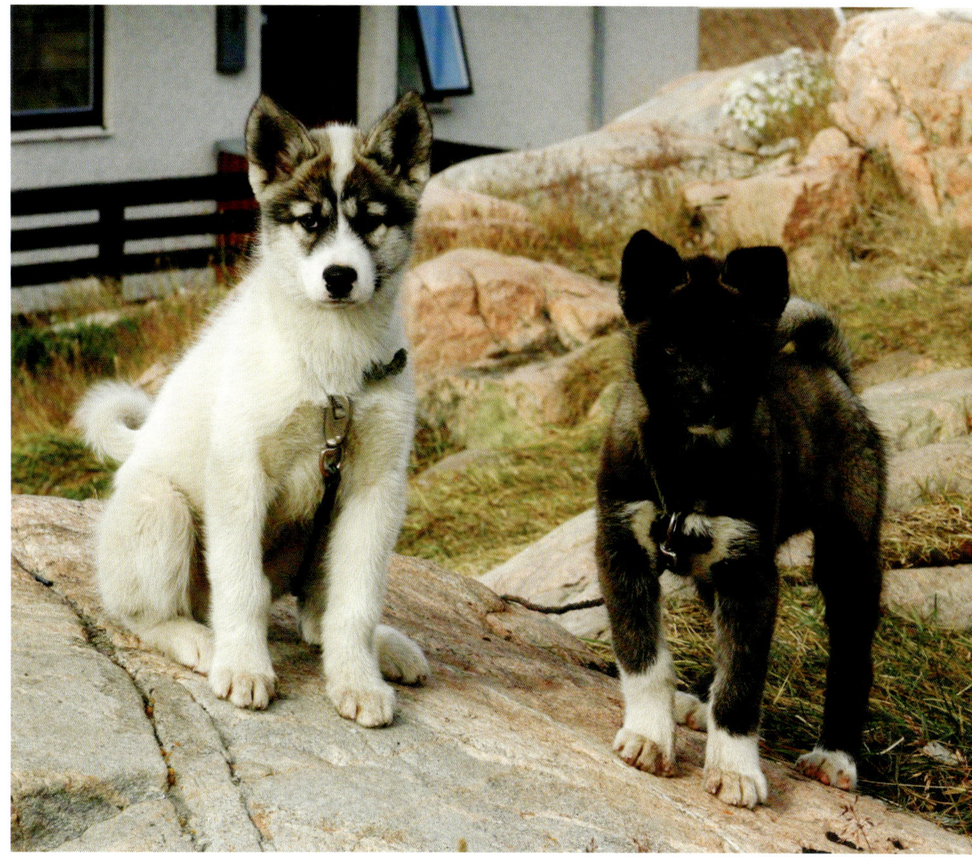

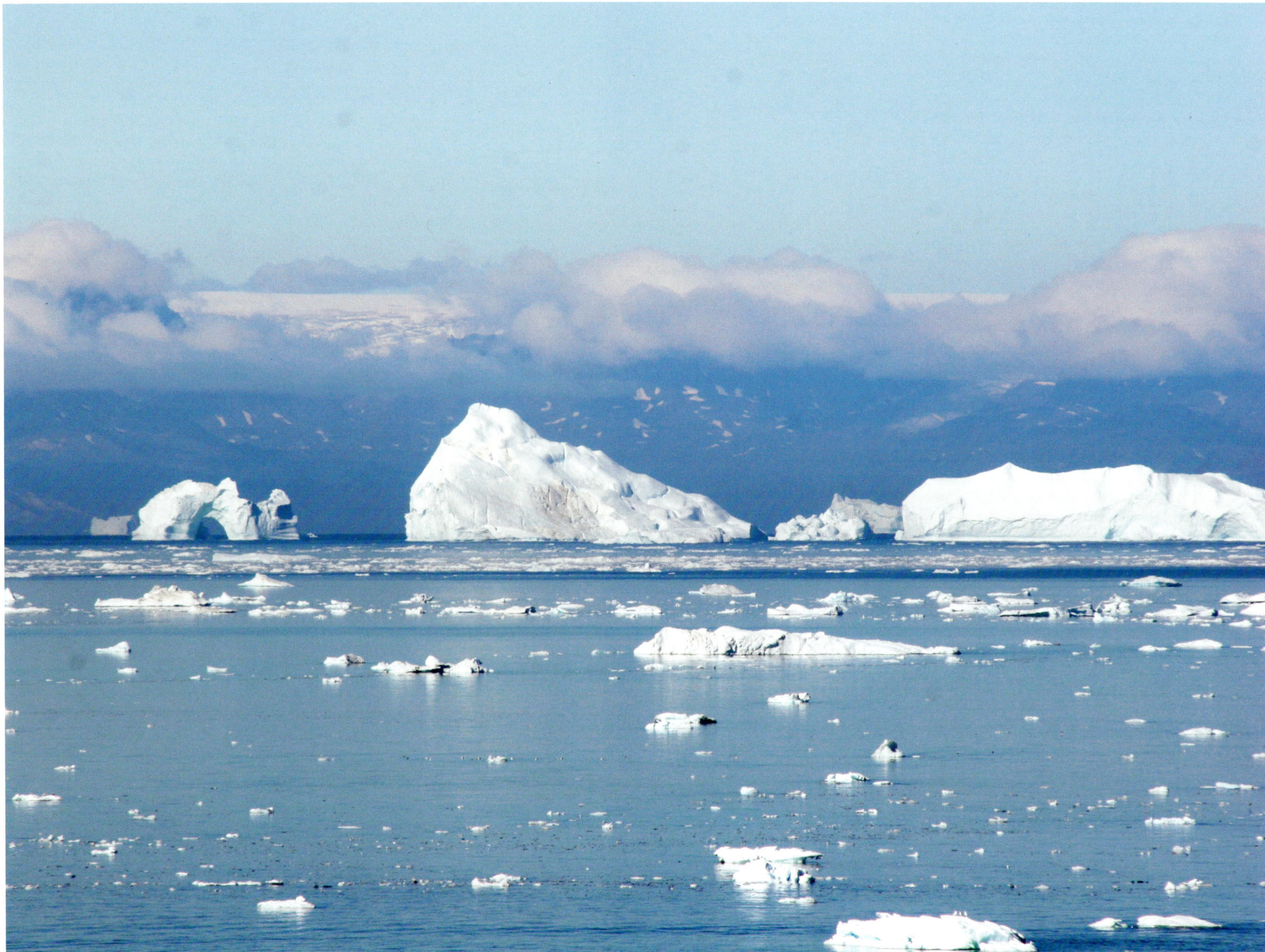

The Pyramid of Disko Bay

Steve

The pyramidal monster that my father calculated to be 175 metres high and over 400 metres long, was far out in the bay, yet it remained as stationary and unchanged as its distant backdrop – Disko Island. The daily ritual of viewing the time-lapse movies from the surveillance cameras confirmed this as the pyramid remained absolutely still while a comical Piccadilly-like circus of smaller bergs zipped this way and that in accelerated time.

I had used up almost one week of precious time-lapse camera coverage on this berg hoping to gauge the mechanisms and rate of berg deterioration in open water. It was after all one of the few grounded bergs upon which wave action may have been expected to occur. Thus it was with painful irony that I removed that particular camera for a short interval on the last evening – only to find upon replacement that our familiar landmark was no more – it had catastrophically collapsed, rolled and broken in what must have been a tectonic scaled incident.

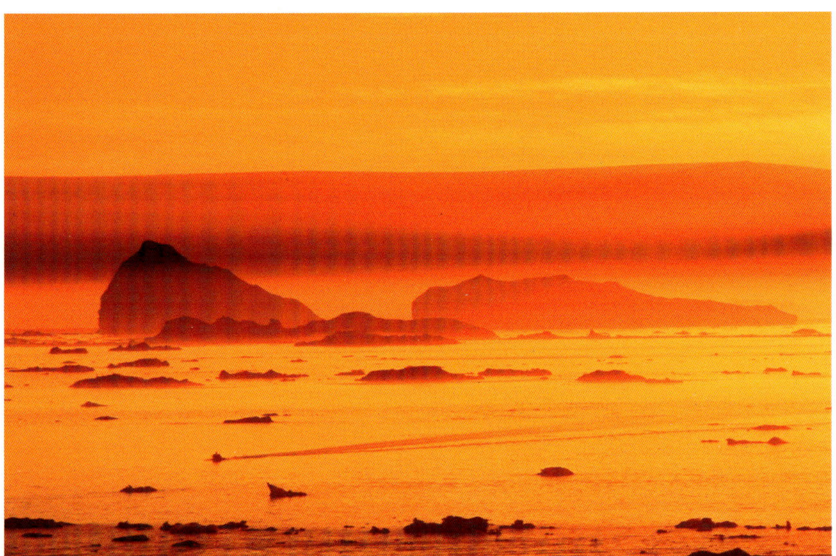

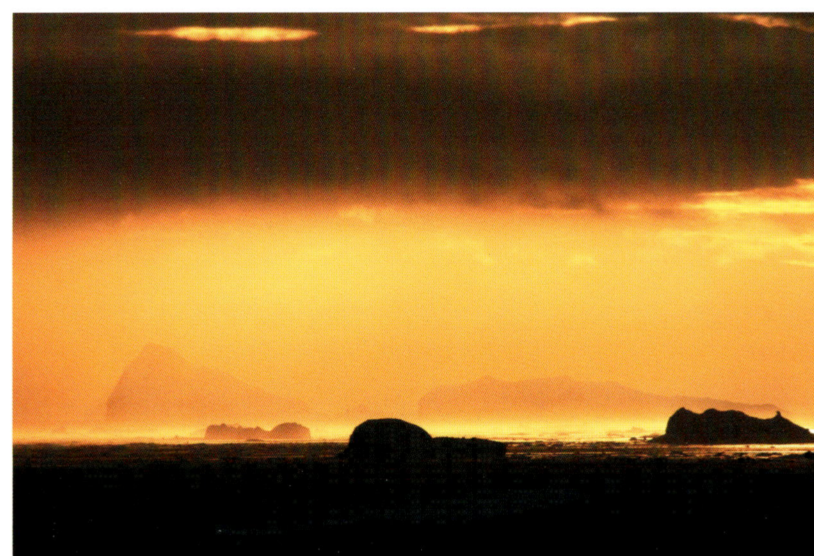

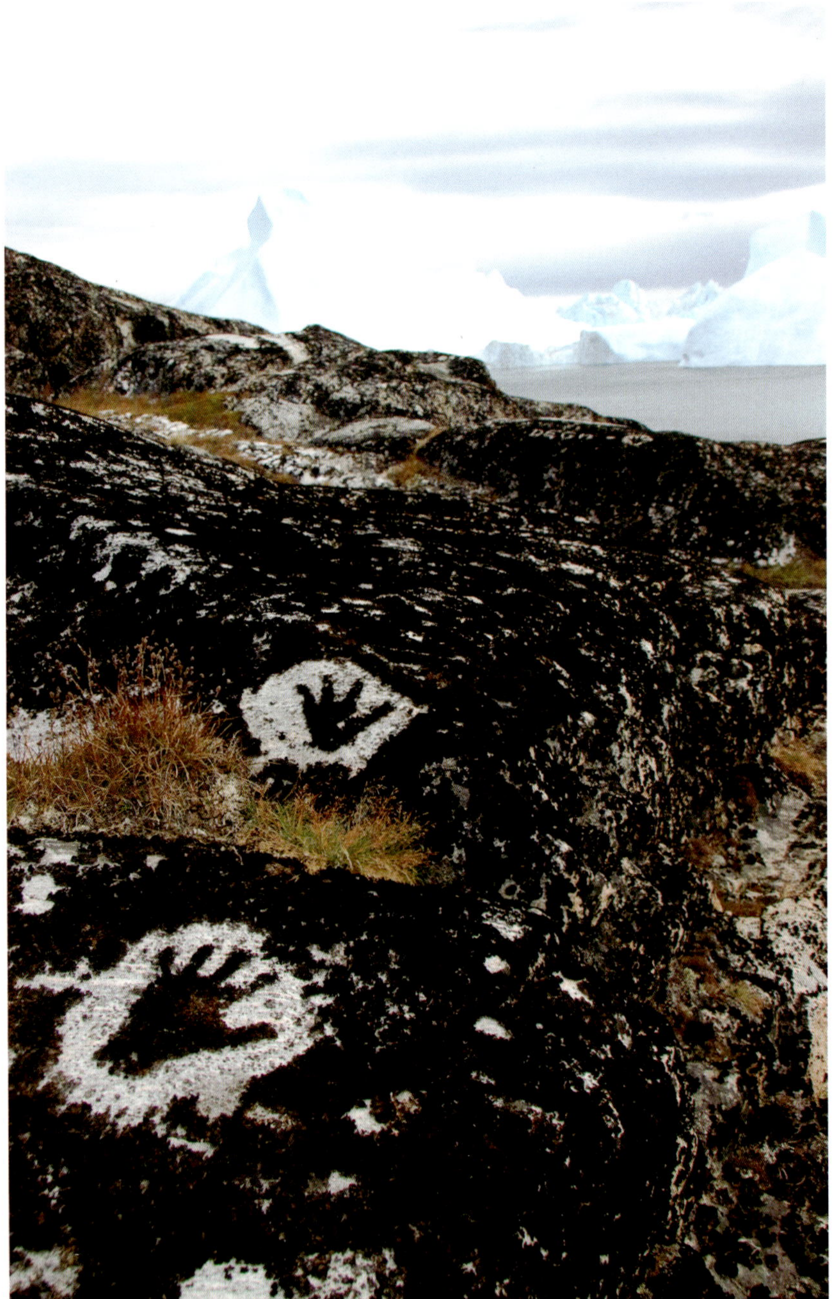

A Message

Steve

FROM THE FRONT

On my last evening in Ilulissat I placed a message in a bottle and tossed it into the massive fjord full of giant icebergs. The thought had occurred to me to do this as I wondered which of the bergs before me would last to have as its final resting place, a cove within a stone's throw of my home in St. John's, Newfoundland.

Two years later and over 2000 km. away I received a call from a trail volunteer that something of interest to me had been found in a bottle on the beach near St. John's. I was stunned by the possibility that my Greenland message might actually have shown up. Unfortunately it turned out to be something else with a faded map on it including Greenland – and the person had linked it to me, having heard mention of my work. So the bottle is still at large, somewhere, and the message, still valid: If found, please contact.

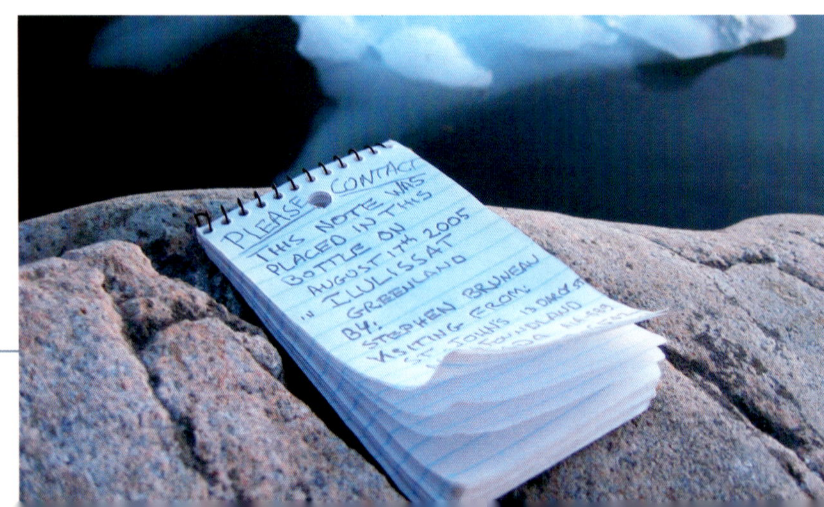

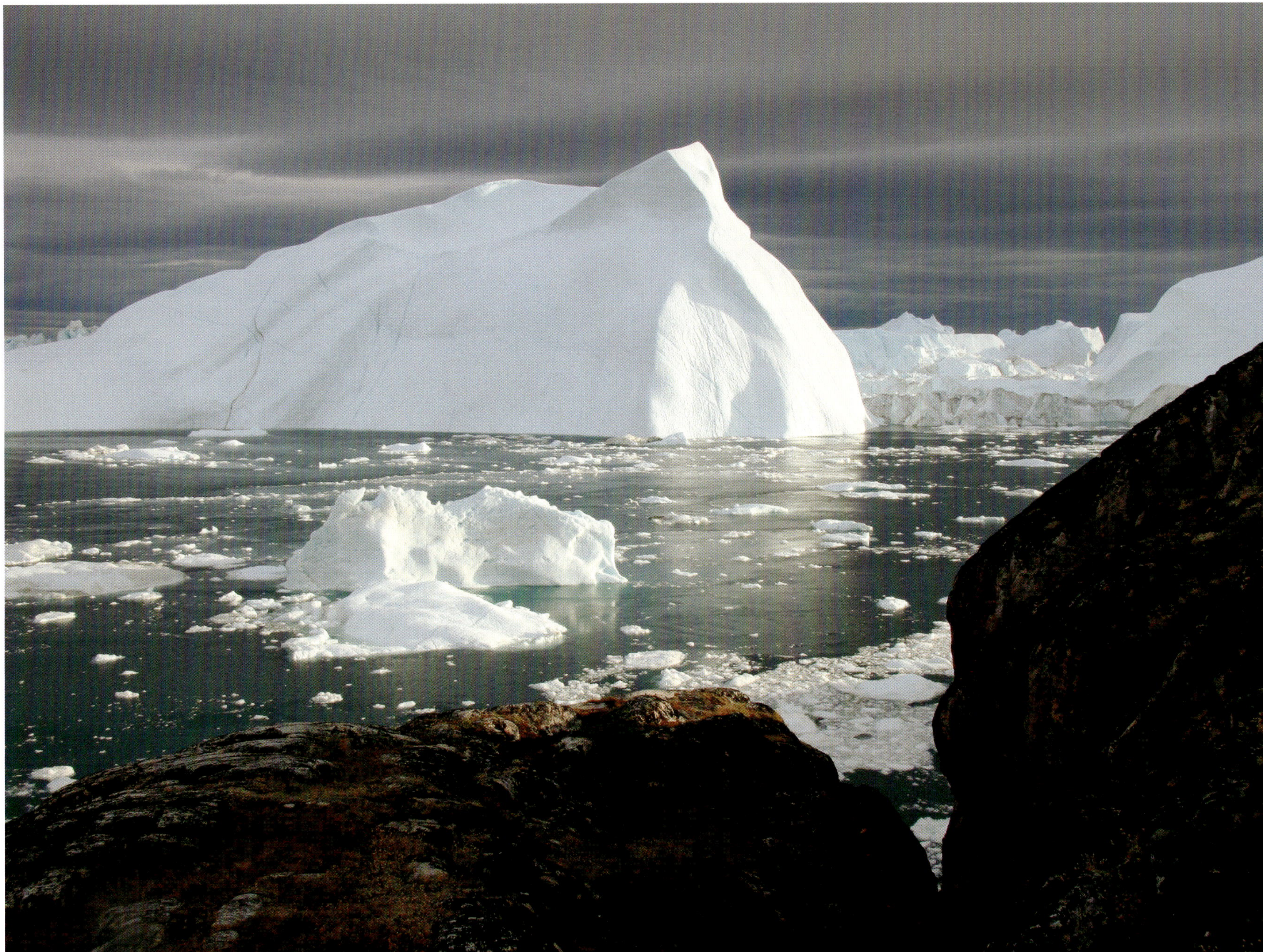

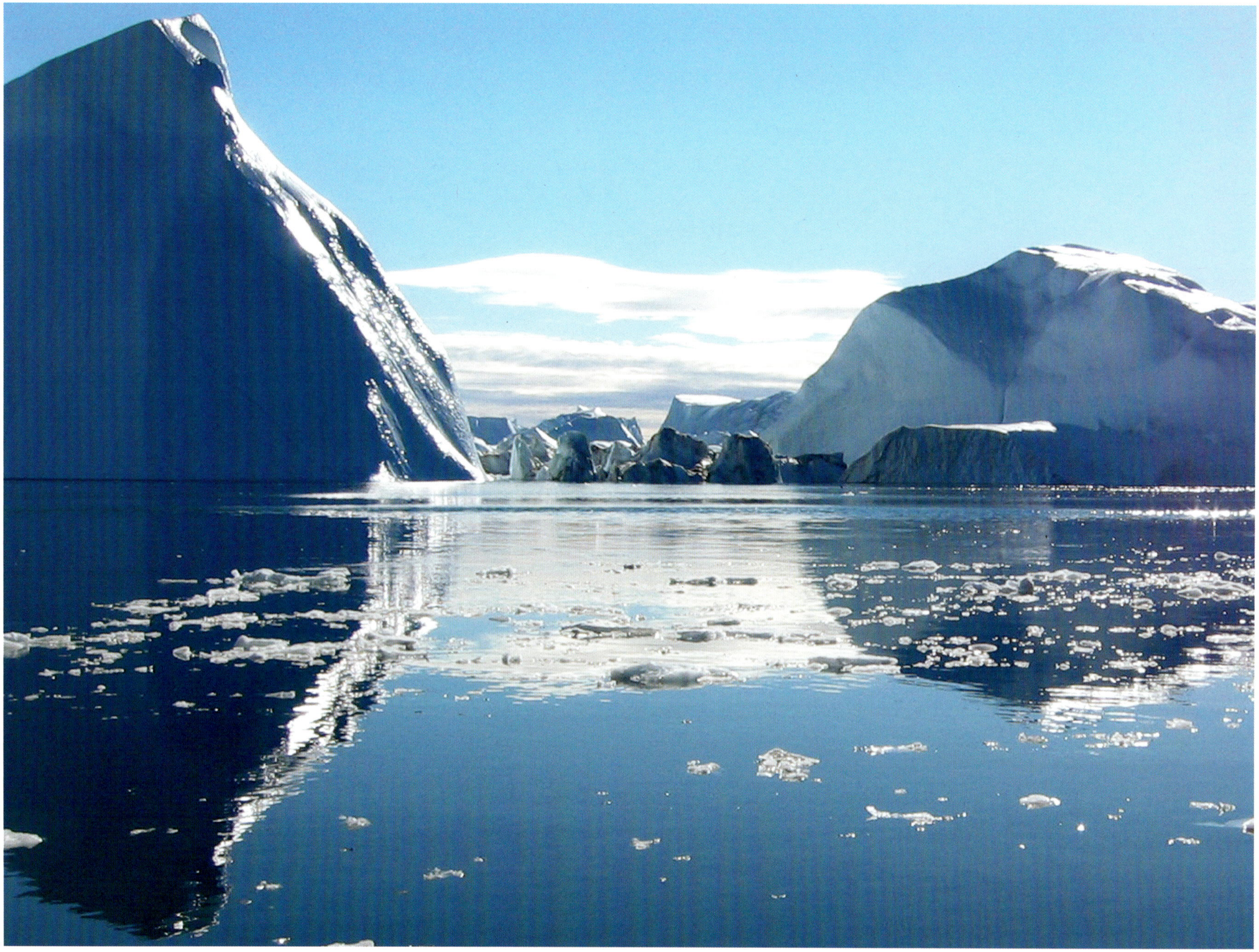

On to Baffin Bay

Steve

Of all the Islands on earth, Greenland, Baffin, and Ellesmere happen to be ranked one, five, and ten respectively in terms of size. Bounded by these huge land masses is Baffin Bay – a vast arctic sea clogged with sea ice for much of the year. It is this Bay that receives the majority of icebergs produced in the Northern Hemisphere, and in late summer and fall it is also in this Bay that most of them melt.

On a research trip to Baffin Island in 1972 my father, Angus, captured pictures of icebergs that had made it through the melting season and overwintered around the coast of that immense land. Safe from melting while trapped in stationary ice, these bergs would soon be driven south with the broken pack ice and for some even farther without it. Though tens of thousands of icebergs are generated annually less than ten per cent will make it as far south as Labrador, and only ten per cent of those will be lucky enough to escape the clutches of the Big Land and the Terrifying Labrador Sea to make it to the fifteenth-largest Island in the world – Newfoundland.

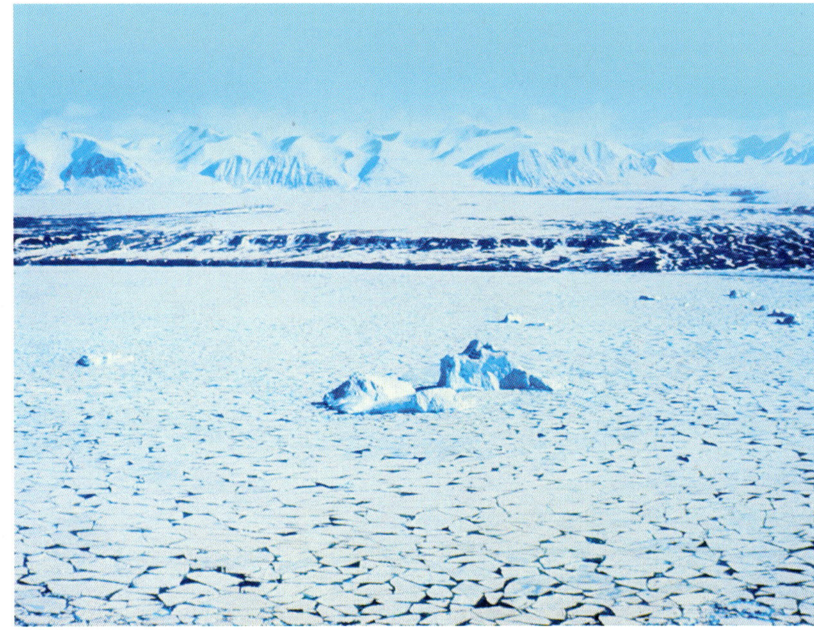

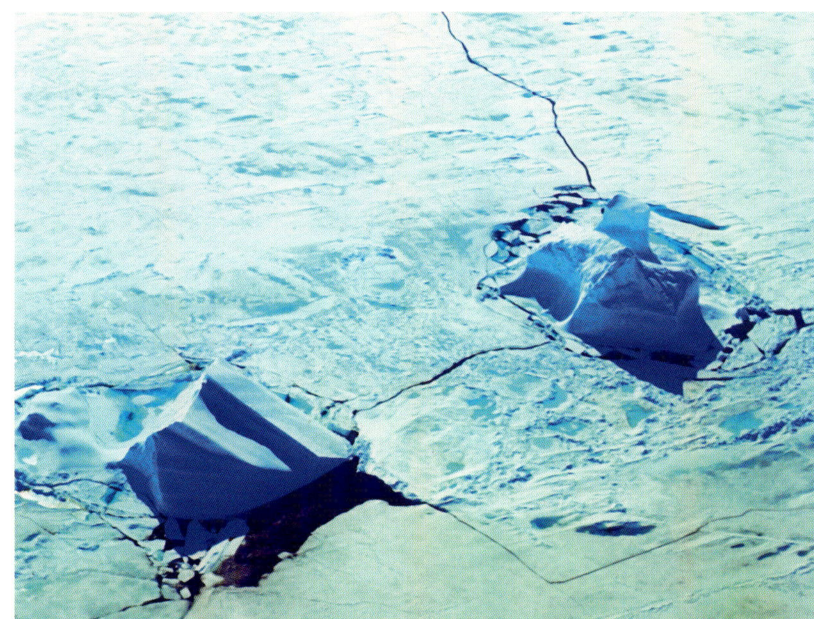

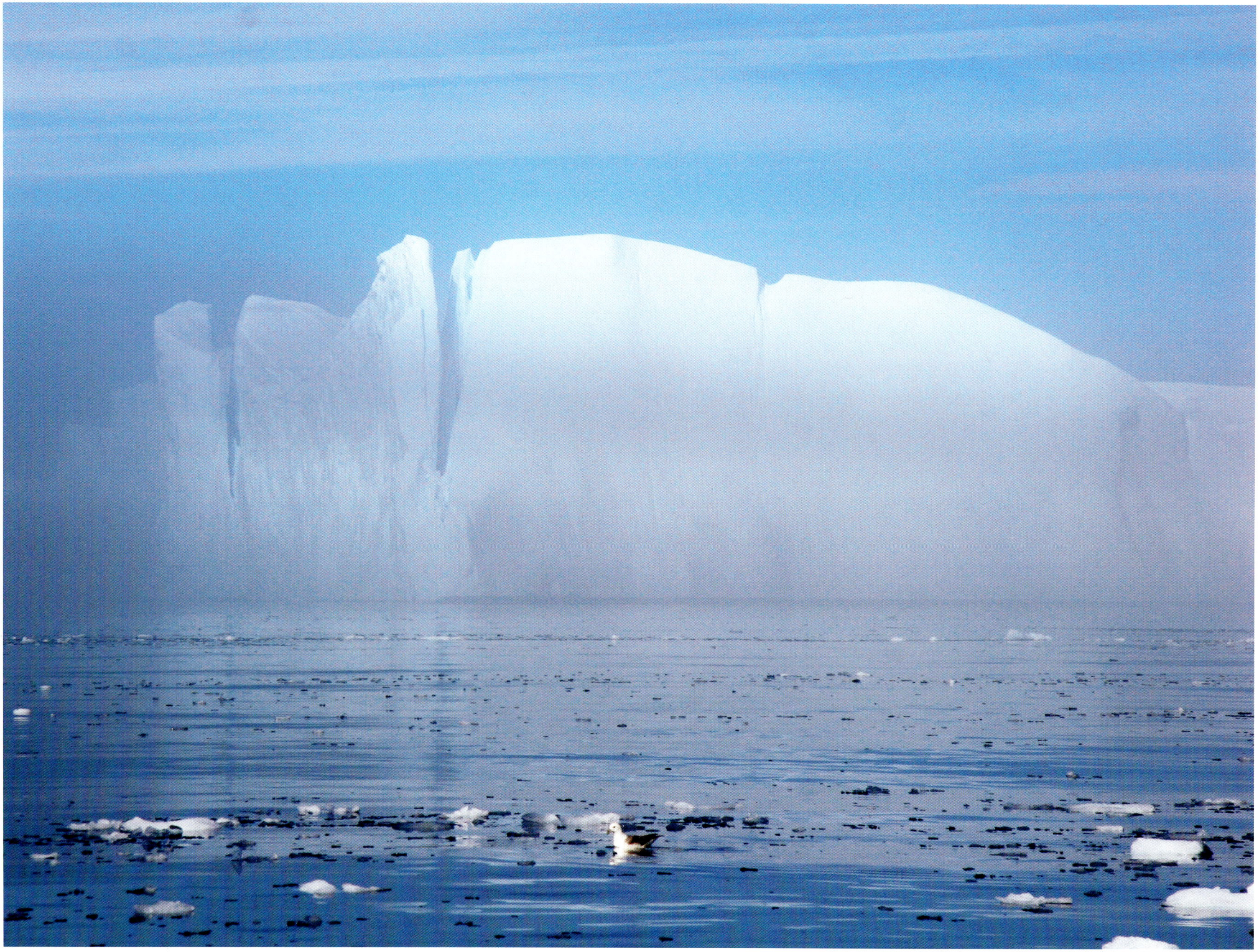

Labrador

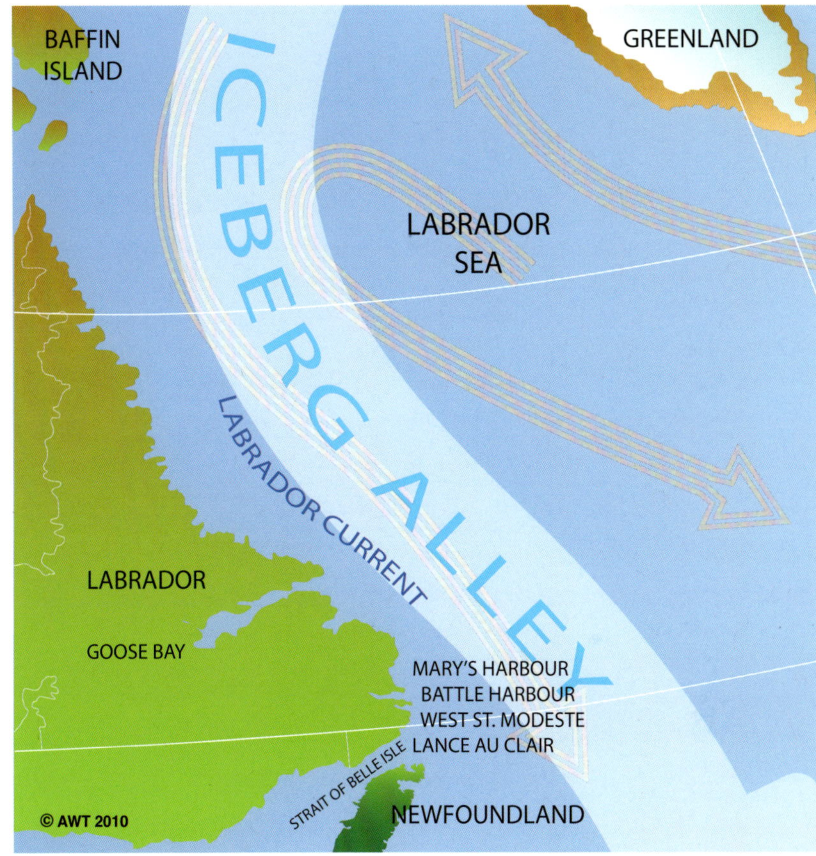
BAFFIN
ISLAND
GREENLAND
ICEBERG ALLEY
LABRADOR
SEA
LABRADOR CURRENT
LABRADOR
GOOSE BAY
MARY'S HARBOUR
BATTLE HARBOUR
WEST ST. MODESTE
LANCE AU CLAIR
STRAIT OF BELLE ISLE
NEWFOUNDLAND
© AWT 2010

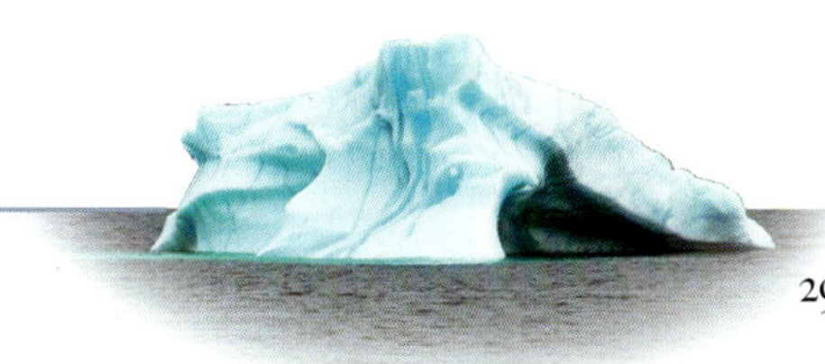

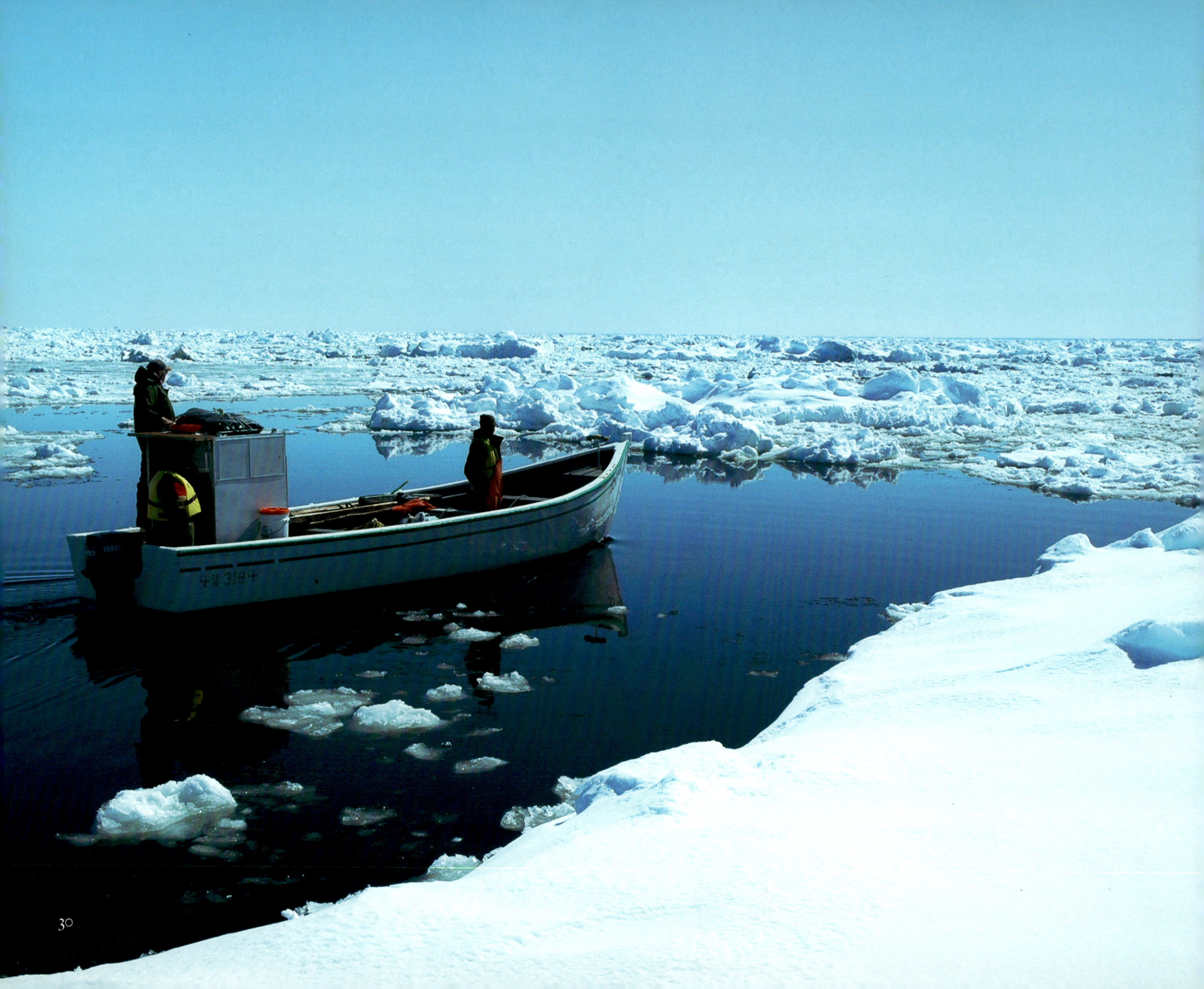

Flotilla of Ice

Kevin

Late arctic pack ice in the Labrador Sea delayed the *Sir Robert Bond*'s first sailing to Goose Bay by three weeks. We were fortunate to have a reservation. En route north, the last leg of the thirty-six-hour passage was most spectacular. Shrouding some icebergs in our path was low-lying cloud. Although banks of fog were a prominent feature, appearing on occasion between fog banks were pockets of clear blue sky and sporadic looming pure whiteness of arctic pack ice.

Although it was summer, standing on deck felt more like early spring . . . the similarities surrounding the *Titanic*'s sinking April 15, 1912, and our current conditions bemused the mind between reality and the surreal. Water calmed by a blanket of arctic pack ice, in concert with each light foggy breeze, emulated the sound of music in the distance. The vessel's vibration, smokestack's diesel effluent, ice-chilled air, scattered shrouds of fog, and towering icebergs left us feeling chilled . . . we could be on the *Titanic!* It was easy to visualize that without radar, a vessel of this size would have little chance of avoiding an iceberg in its imminent path.

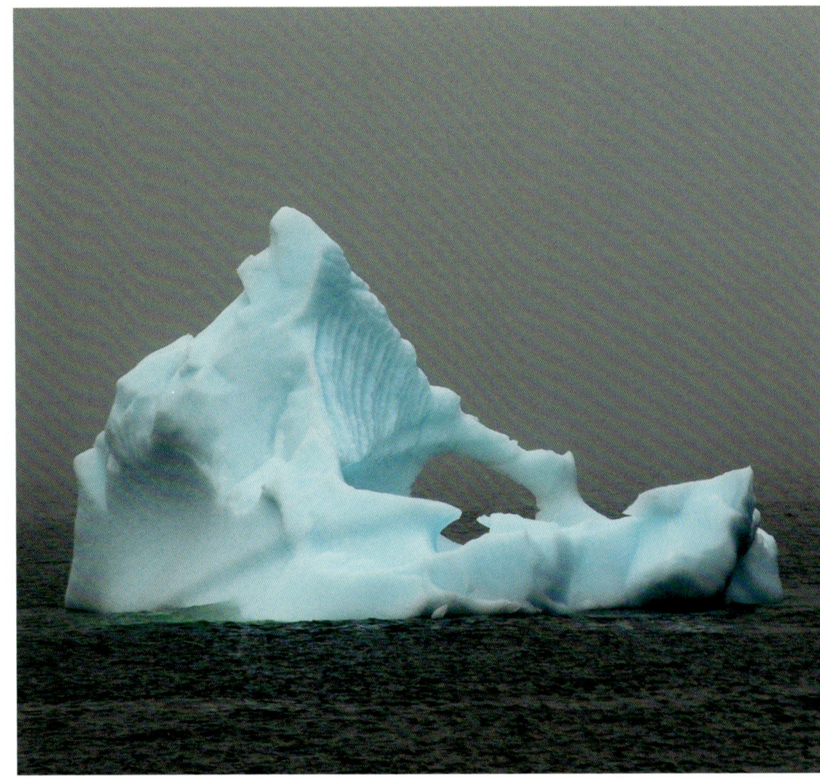

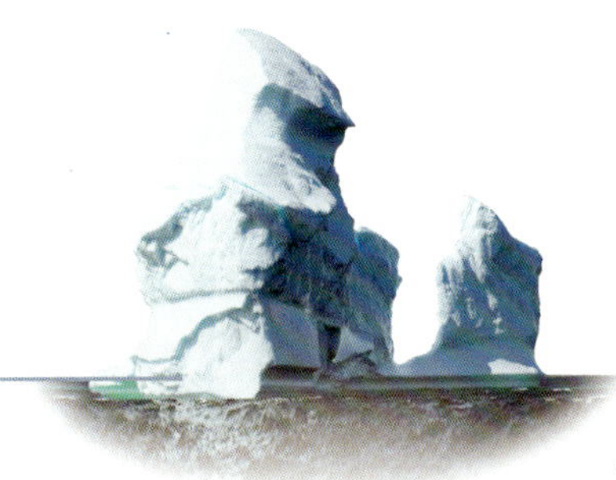

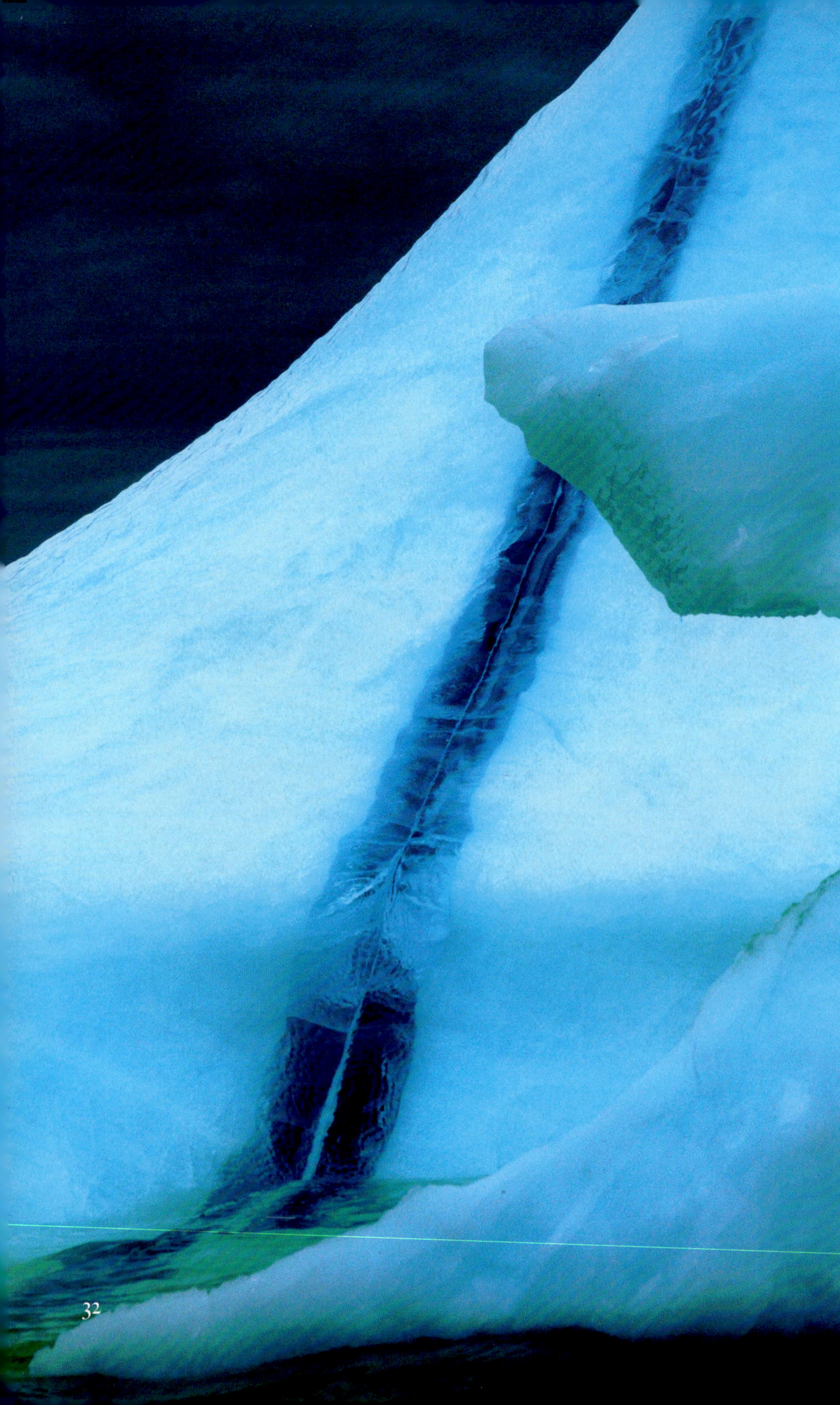

Feathers & Fissures

Kevin

Feathers of ice reaching skyward filtered in late afternoon light. Deep-set fissures parted these feathers of ice like rivulets on a steep-sloped mountainside born from the scouring of a passing glacier. The only difference being these fissures and feathers of ice were shaped with such perfection. They were more like the vision created by a master sculptor or an artist's brush not normally shared with a mortal like me. I knew being here to experience this was a special privilege.

What had created such a masterpiece? They were too smooth to be scoured by rock. No, it couldn't have been the sun or something above the surface. Whatever it was, it happened underwater. Aha! Maybe it was tiny bubbles of trapped air being released from the ice and rising like the carbon dioxide in a carbonated drink. On land, water runs downhill creating rivulets, rivers, and streams. Perhaps this was bold evidence of air rising, caressing the ice from which it was born.

If this was the explanation, it was even more amazing that tiny bubbles of air could create such deep fissures in solid ice.

P.S. *On return home check cause of fissures with iceberg expert Dr. Stephen Bruneau!*

Labrador Coast:
Mary's Harbour

From the distance the Mary's Harbour iceberg appeared small and featureless, but something in my gut urged us to have a look. On closer inspection, we were awestruck for this tiny iceberg revealed features of "feathers and fissures" I had never seen before.

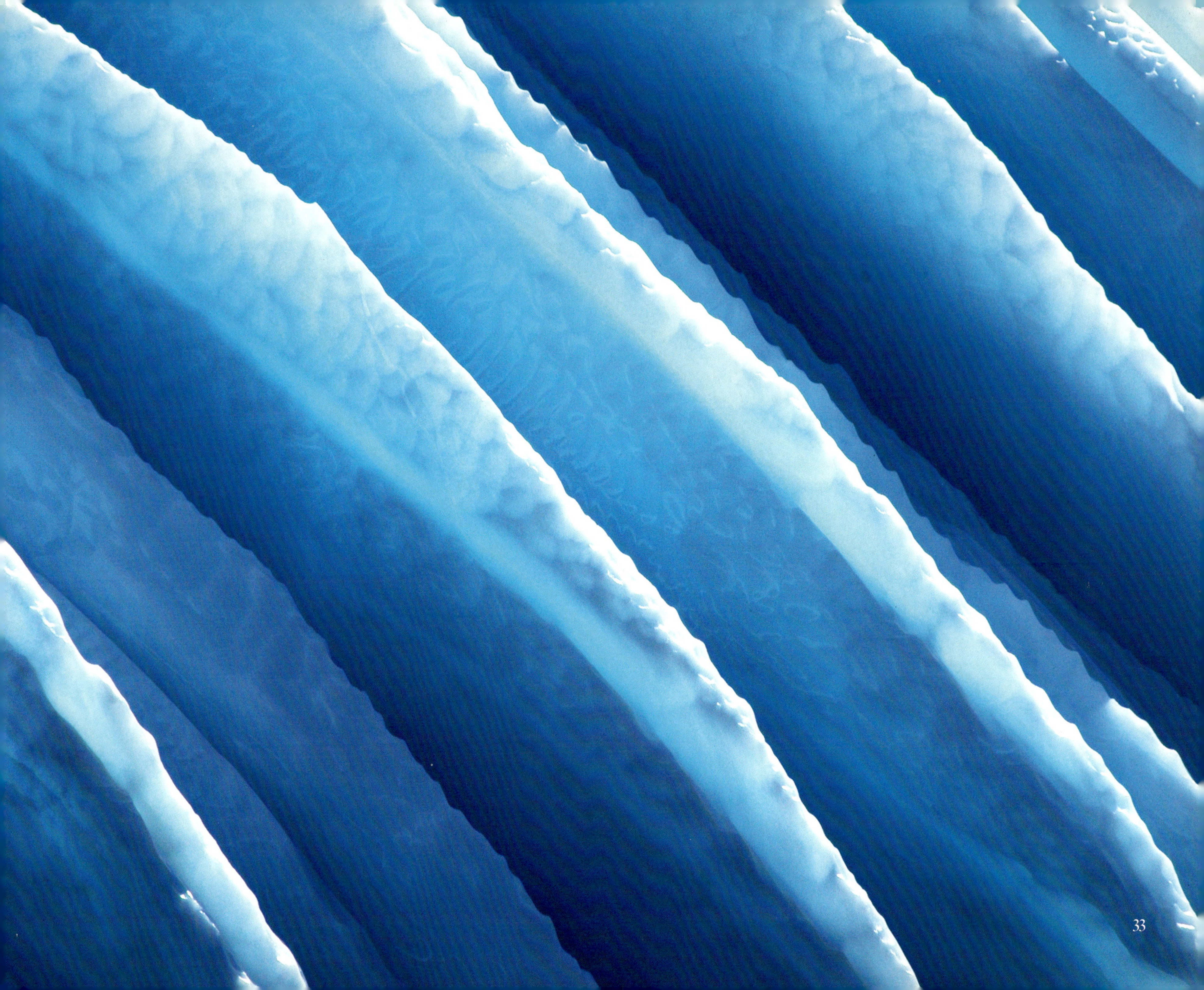

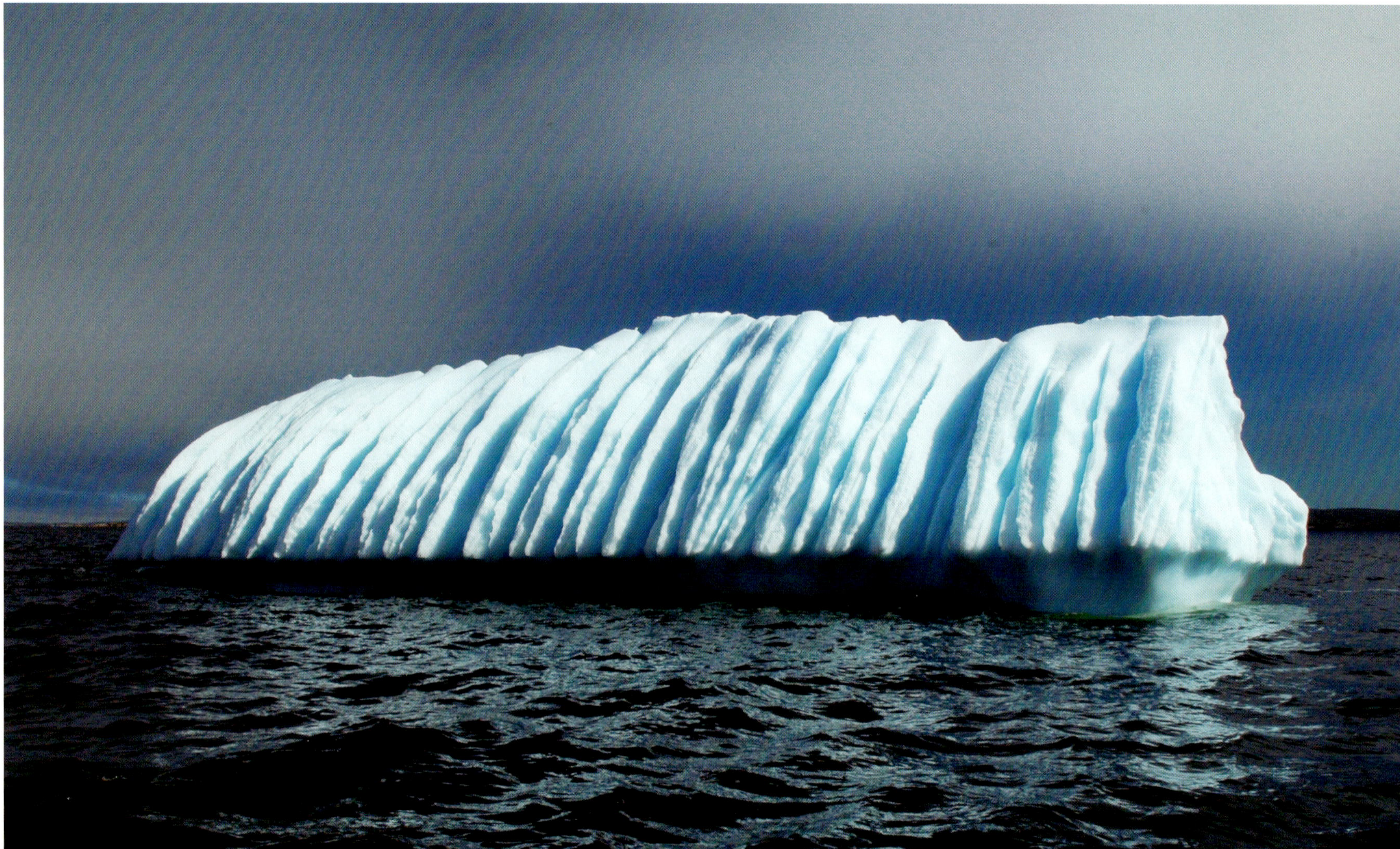

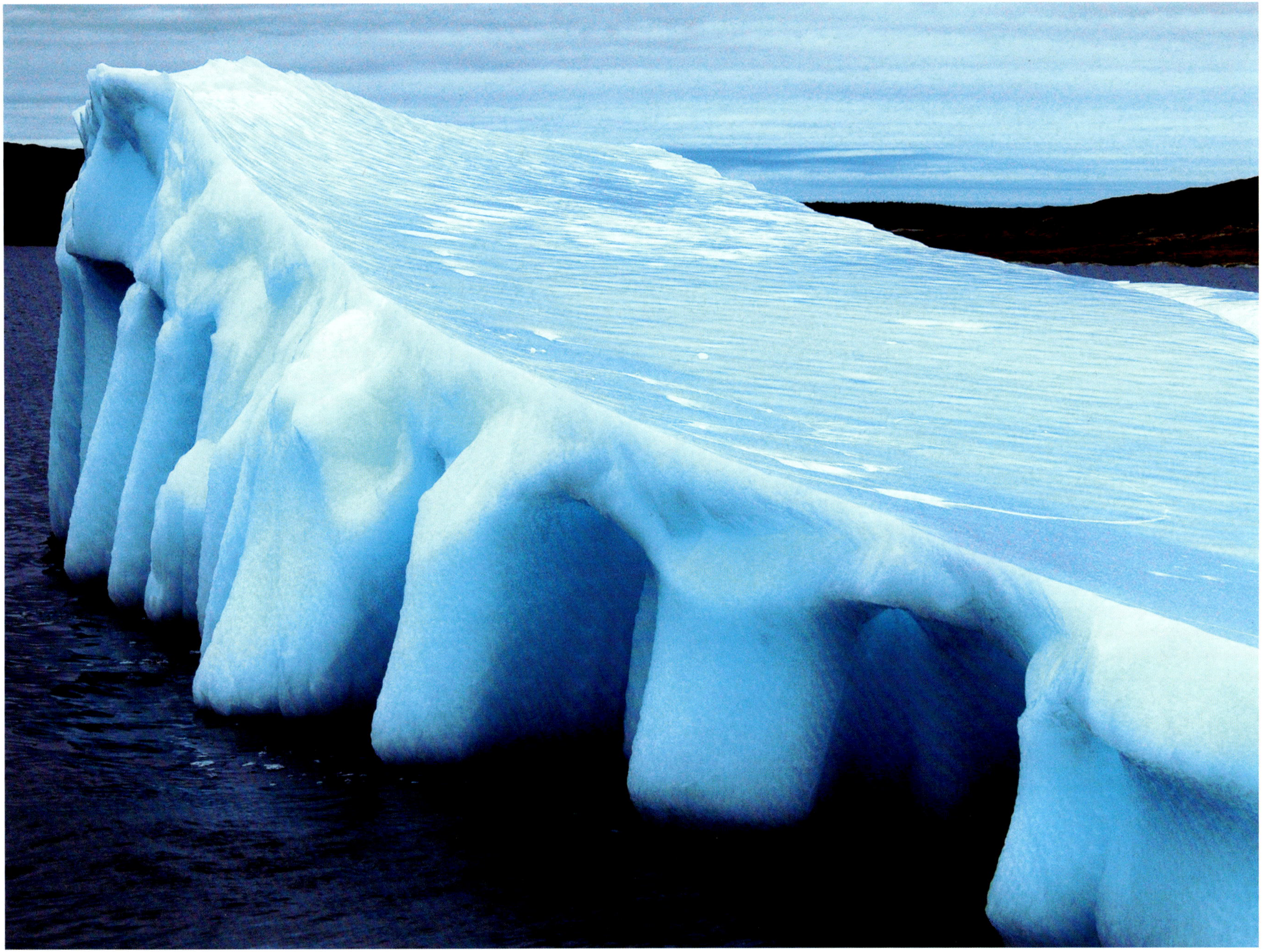

The Great Labrador Gale

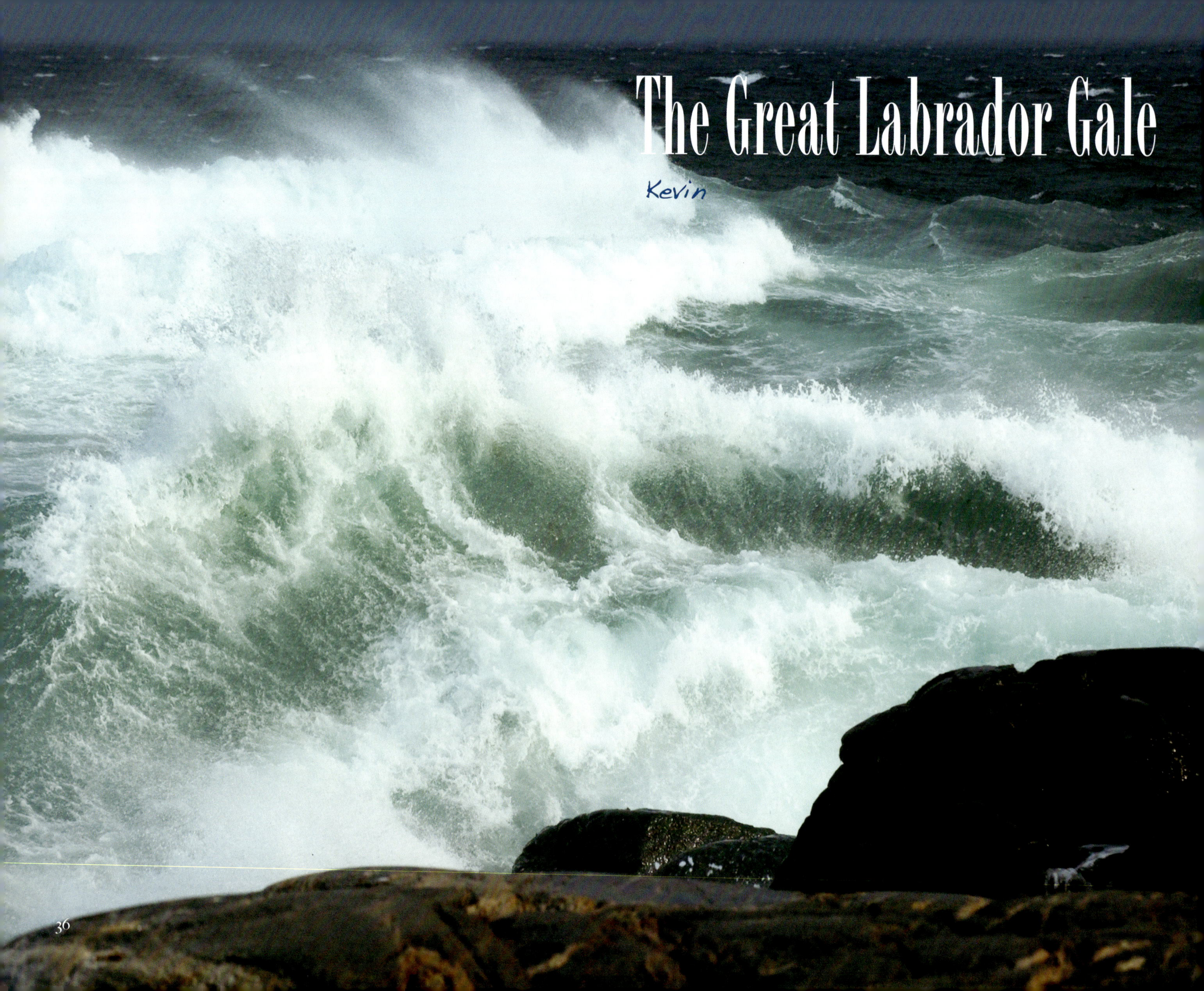

Earlier today we faced a "Labrador Gale" that brought stinging seas, brine-soaked gusting winds, and a myriad of icebergs to Battle Harbour Island, our place of reprieve. Escalating north-easterly winds stalled all marine travel by mid-morning including us in our fragile kayaks. Soothing our egos were the four sixty-foot fishing trawlers who refused to set sail because of the wind we narrowly escaped.

At the height of the storm, raindrops at the horizontal split on impact with our high-tech rain gear fluttering in the wind. From the windswept promontory where we stood, we watched (what we guessed) a two-hundred-thousand-tonne iceberg push through the southern tickle spinning like a merry-go-round, its roots obviously glancing a gnarly seabed.

Within spitting distance (with the wind) from our observation post laid the remnants of a plane wreck, an obvious reminder of nature's forbearance.

June 29, 2000 AD
Battle Harbour;
Coastal Labrador
On October 10, 1885, the "Great Labrador Gale" struck the Labrador coast. According to the Newfoundland Geological Survey, "A long period of stormy weather eventually produced 120 mph winds and a 'tidal wave' causing considerable damage along the coast and 300 direct deaths."

Roger de Bussey-Rabutin said "Absence is to love as wind is to fire; it extinguishes the small and kindles the great." The June 29 winds of Battle Harbour kindled a lifelong heartfelt intimacy with an isolated destination of choice enriched by the armada of icebergs blown ashore and gracing our presence.

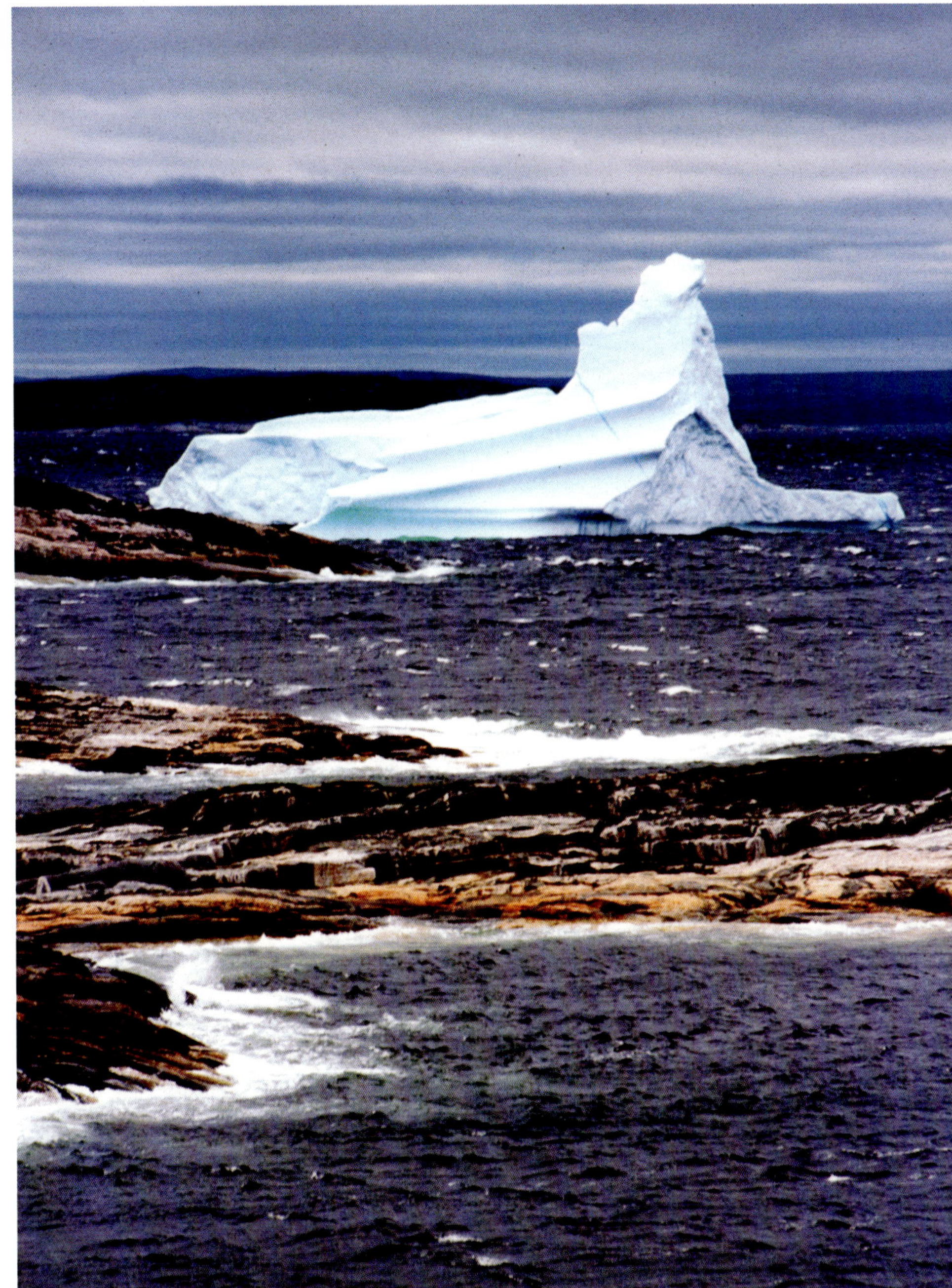

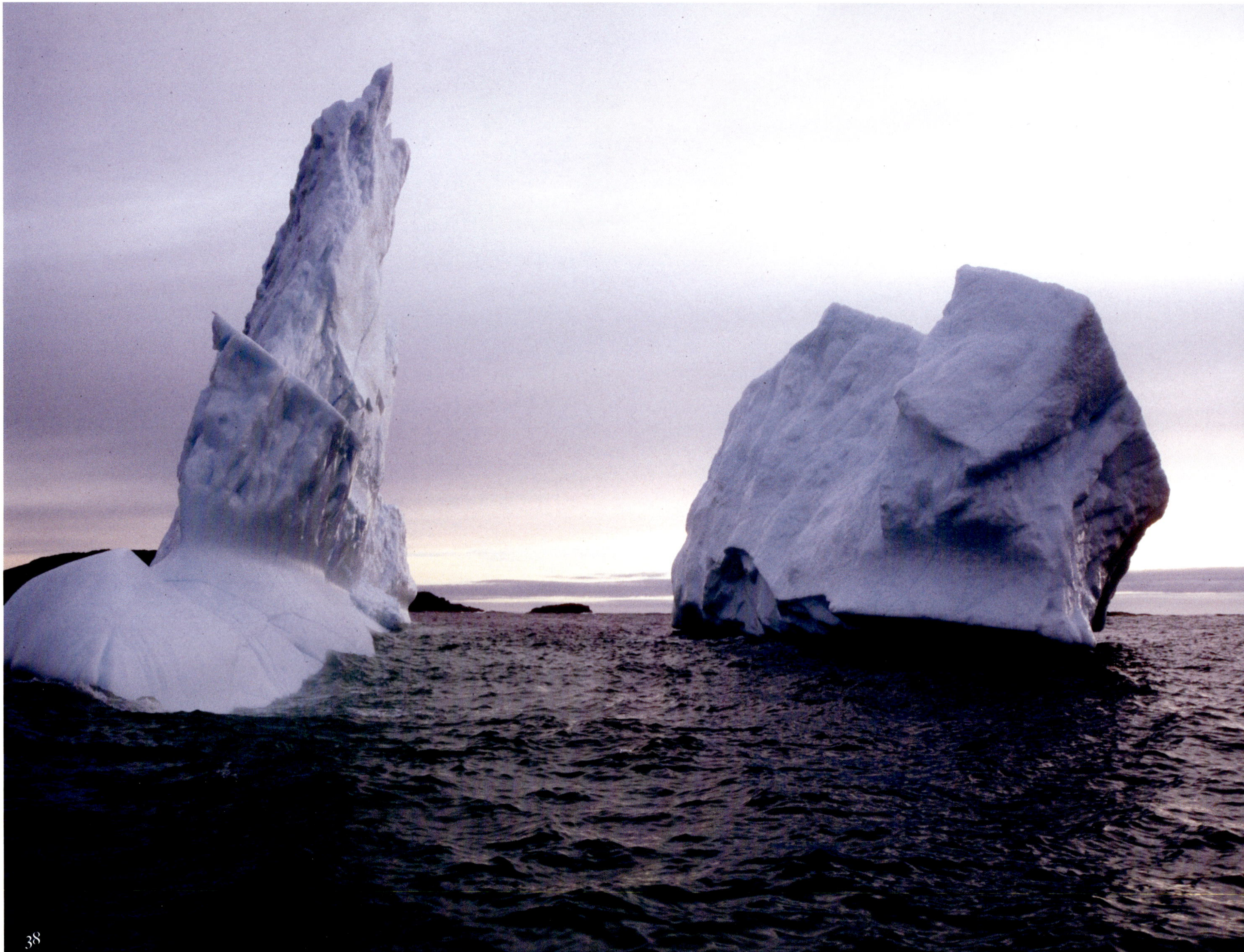

Morning after the Great Gale

Kevin

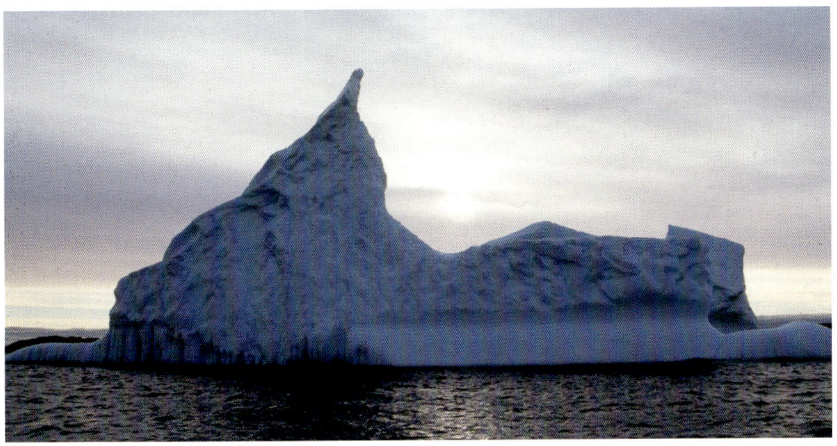

June 30, 2000 AD
Battle Harbour,
Coastal Labrador

Awakened by the absence of wind, we climbed the same windswept promontory of Battle Harbour Island, following the route of so many others before us, including Sir Robert Perry, before and after his journey to the North Pole.

Two days ago, this vista revealed only a couple of icebergs. Today, however, icebergs floated in surrounding waters with abundance like white clouds against contrasting skies. We stopped counting at thirty-four, millions of tonnes of ice blown shoreward in one day: truly, a powerful reminder of the strength of the sea.

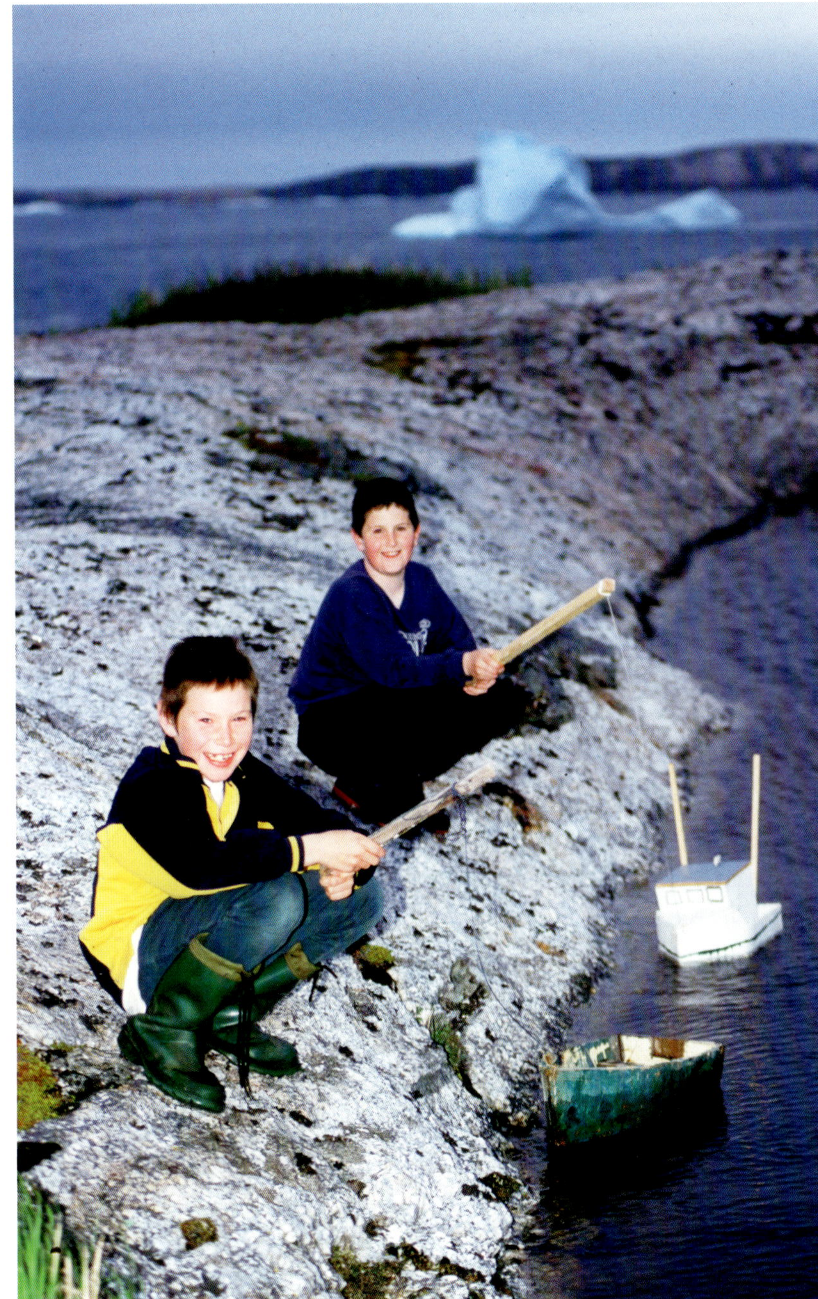

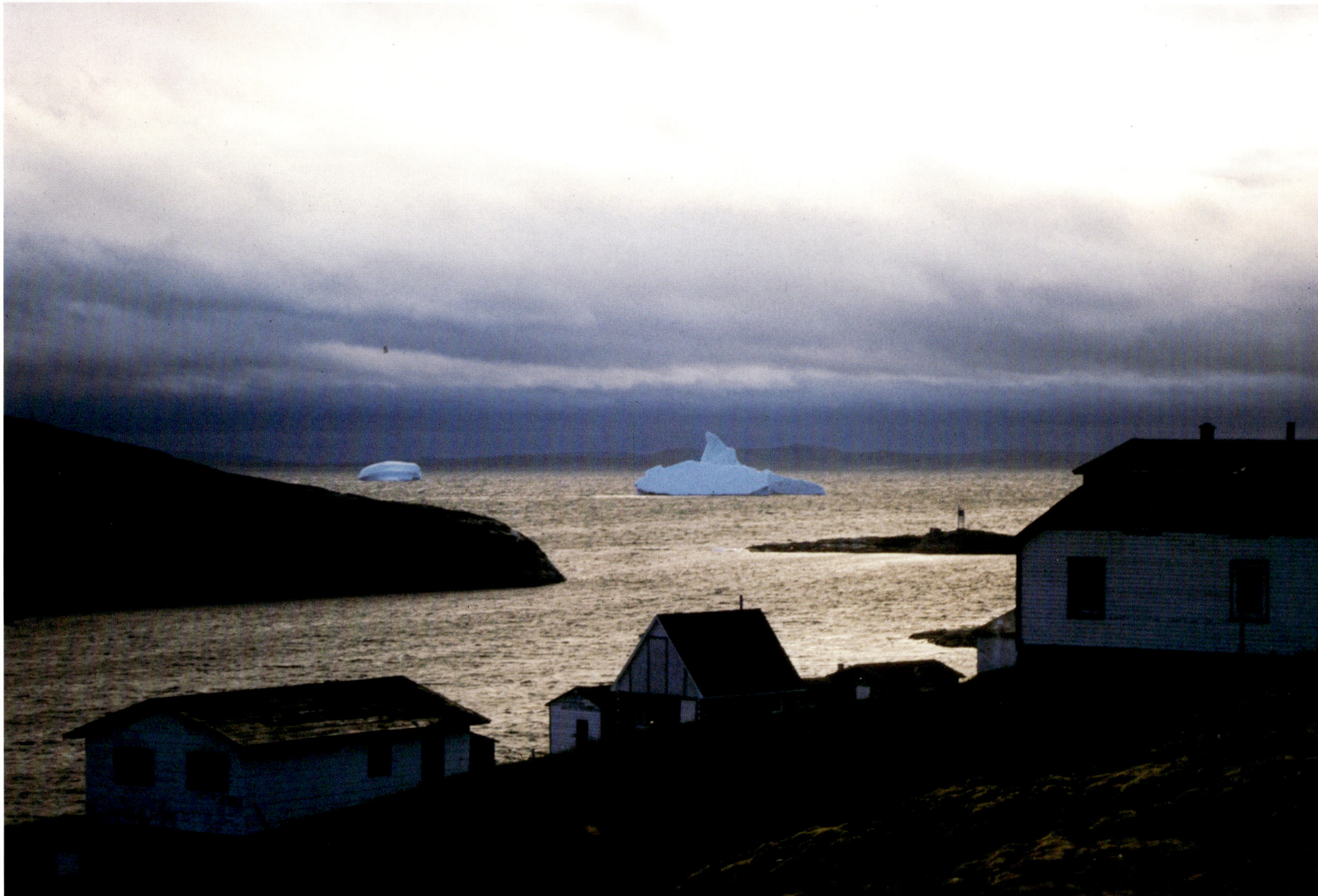

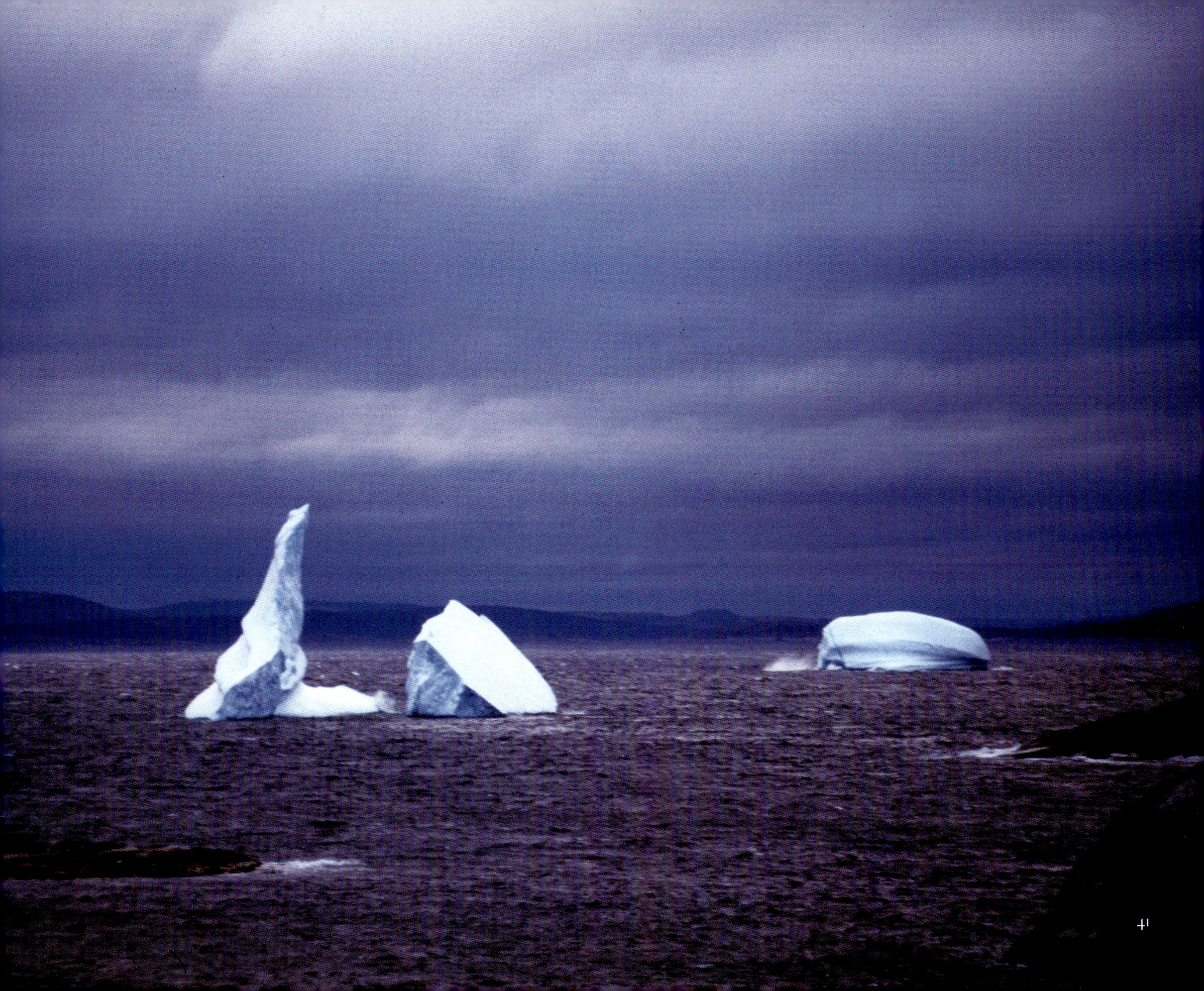

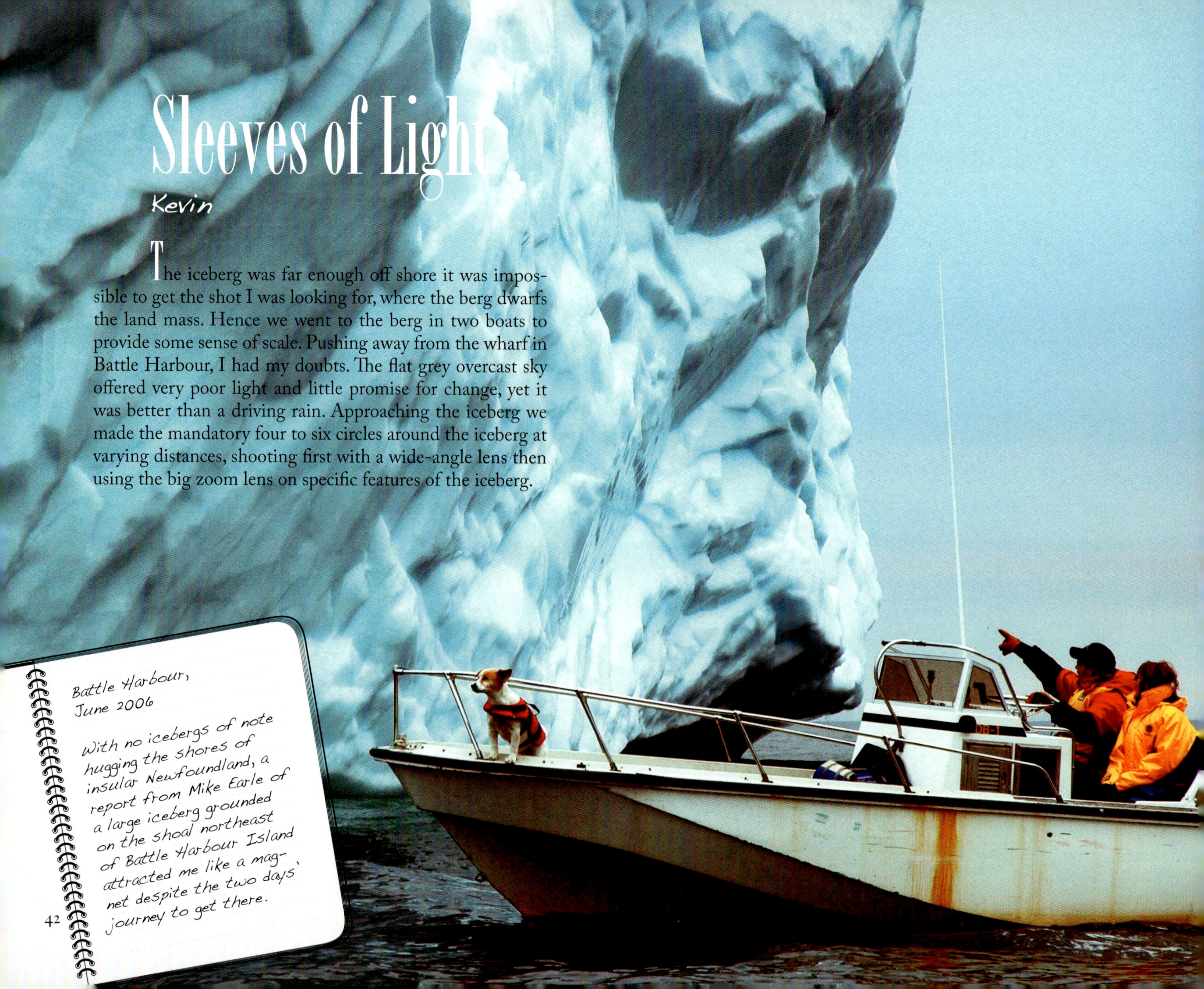

Sleeves of Light

Kevin

The iceberg was far enough off shore it was impossible to get the shot I was looking for, where the berg dwarfs the land mass. Hence we went to the berg in two boats to provide some sense of scale. Pushing away from the wharf in Battle Harbour, I had my doubts. The flat grey overcast sky offered very poor light and little promise for change, yet it was better than a driving rain. Approaching the iceberg we made the mandatory four to six circles around the iceberg at varying distances, shooting first with a wide-angle lens then using the big zoom lens on specific features of the iceberg.

Battle Harbour,
June 2006

With no icebergs of note hugging the shores of insular Newfoundland, a report from Mike Earle of a large iceberg grounded on the shoal northeast of Battle Harbour Island attracted me like a magnet despite the two days' journey to get there.

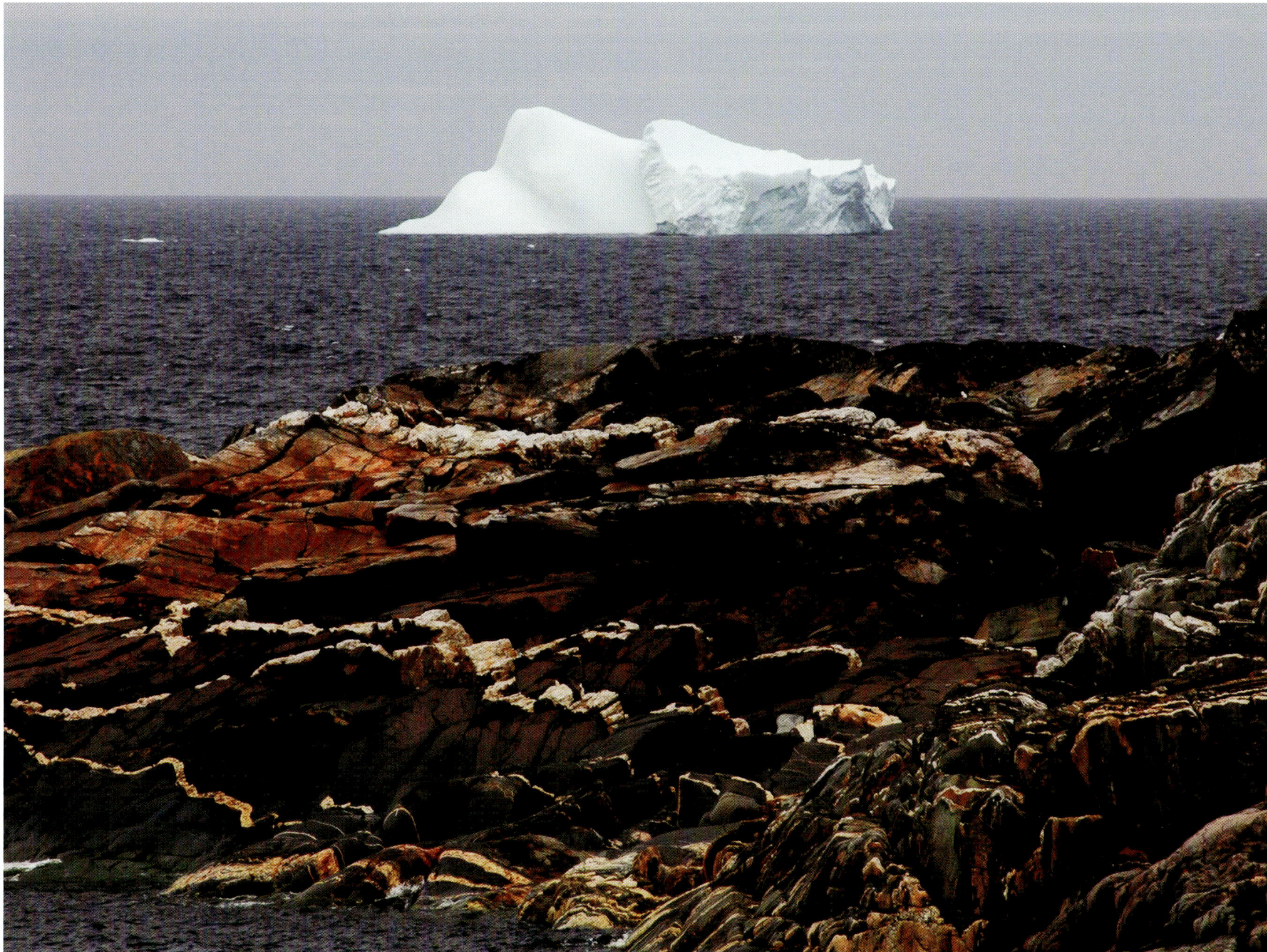

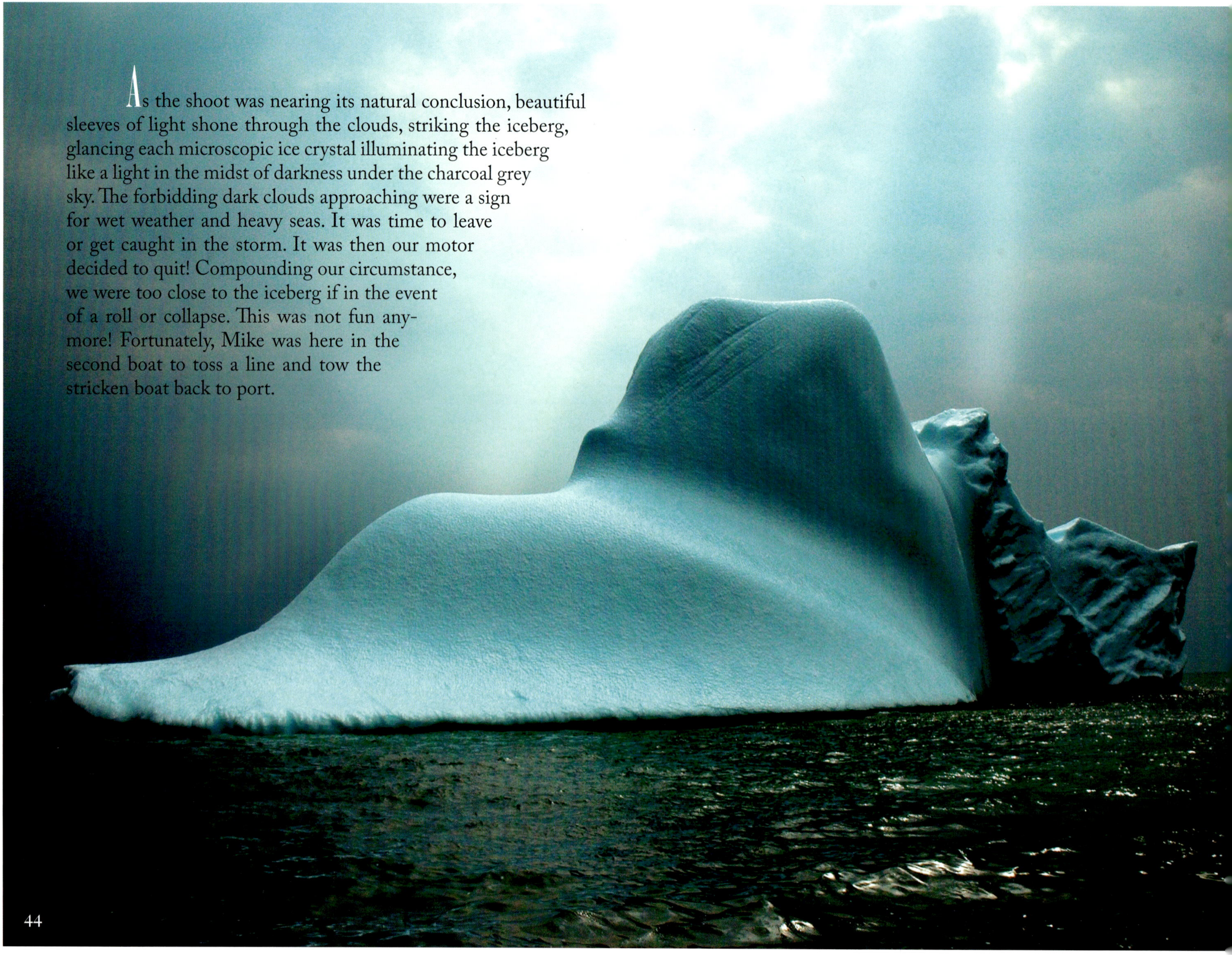

As the shoot was nearing its natural conclusion, beautiful sleeves of light shone through the clouds, striking the iceberg, glancing each microscopic ice crystal illuminating the iceberg like a light in the midst of darkness under the charcoal grey sky. The forbidding dark clouds approaching were a sign for wet weather and heavy seas. It was time to leave or get caught in the storm. It was then our motor decided to quit! Compounding our circumstance, we were too close to the iceberg if in the event of a roll or collapse. This was not fun anymore! Fortunately, Mike was here in the second boat to toss a line and tow the stricken boat back to port.

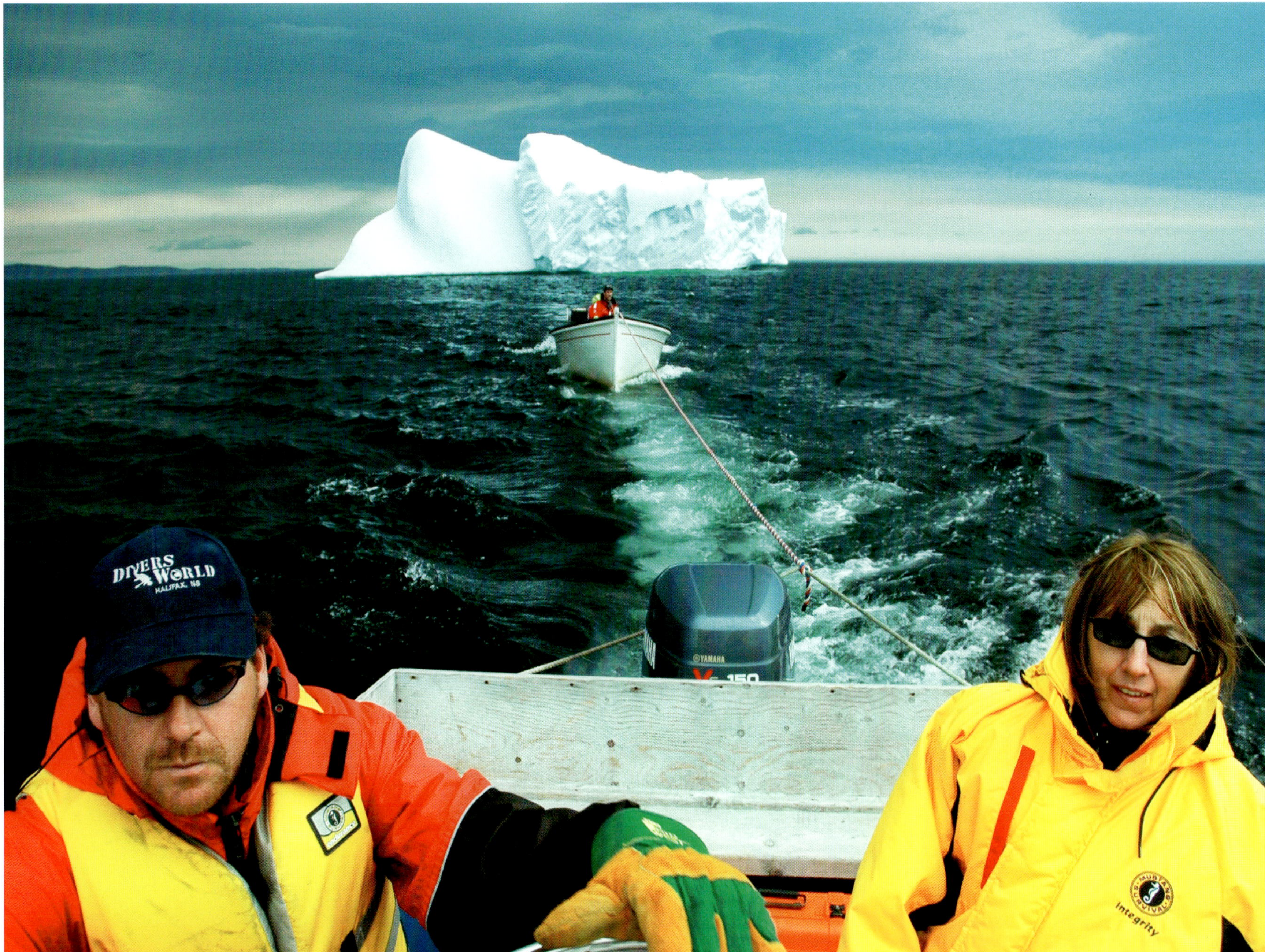
DIVERS WORLD
HALIFAX, NS
YAMAHA
Integrity

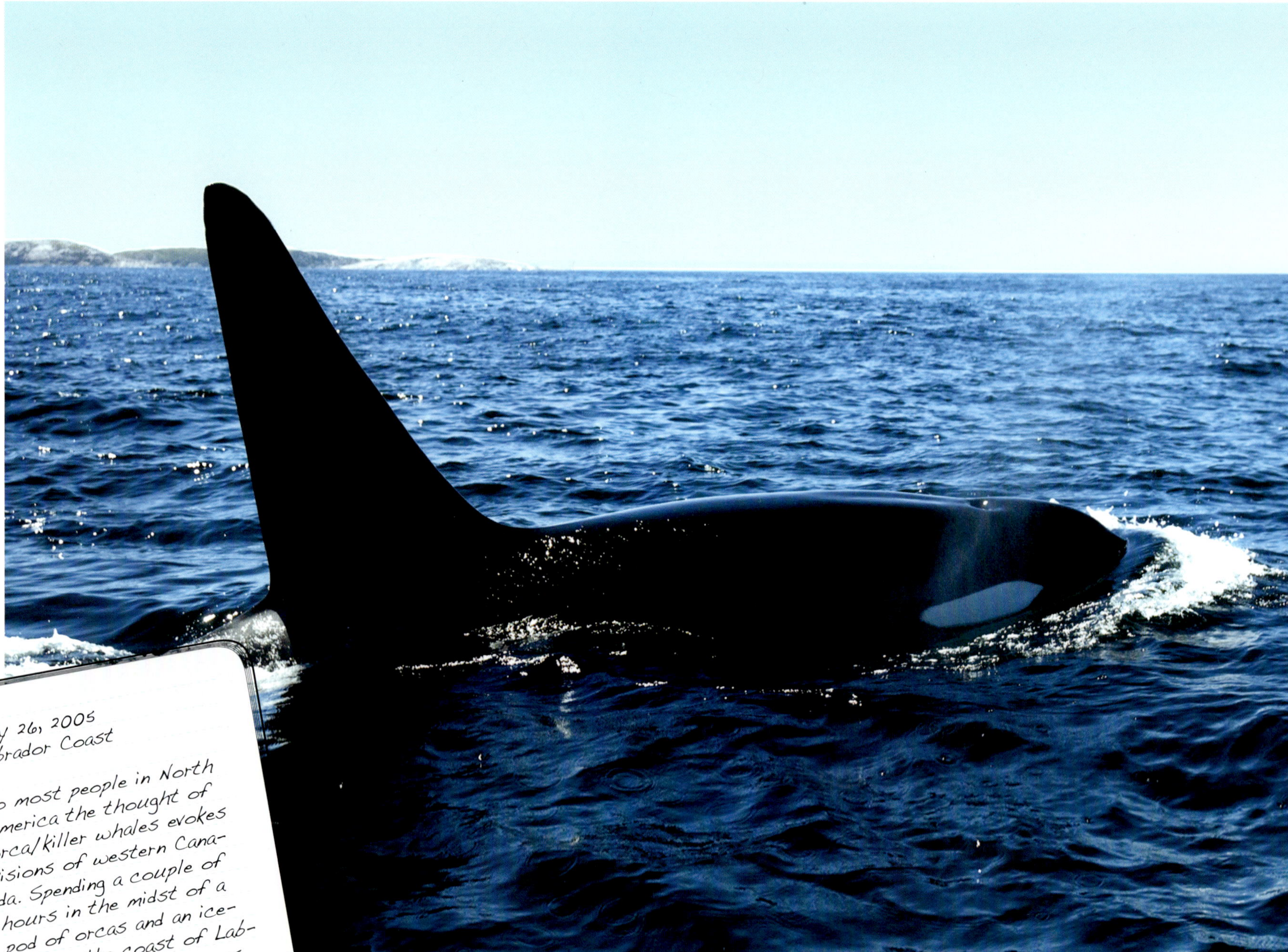

July 26, 2005
Labrador Coast

To most people in North America the thought of orca/killer whales evokes visions of western Canada. Spending a couple of hours in the midst of a pod of orcas and an iceberg on the coast of Labrador is an experience as rare as a precious jewel.

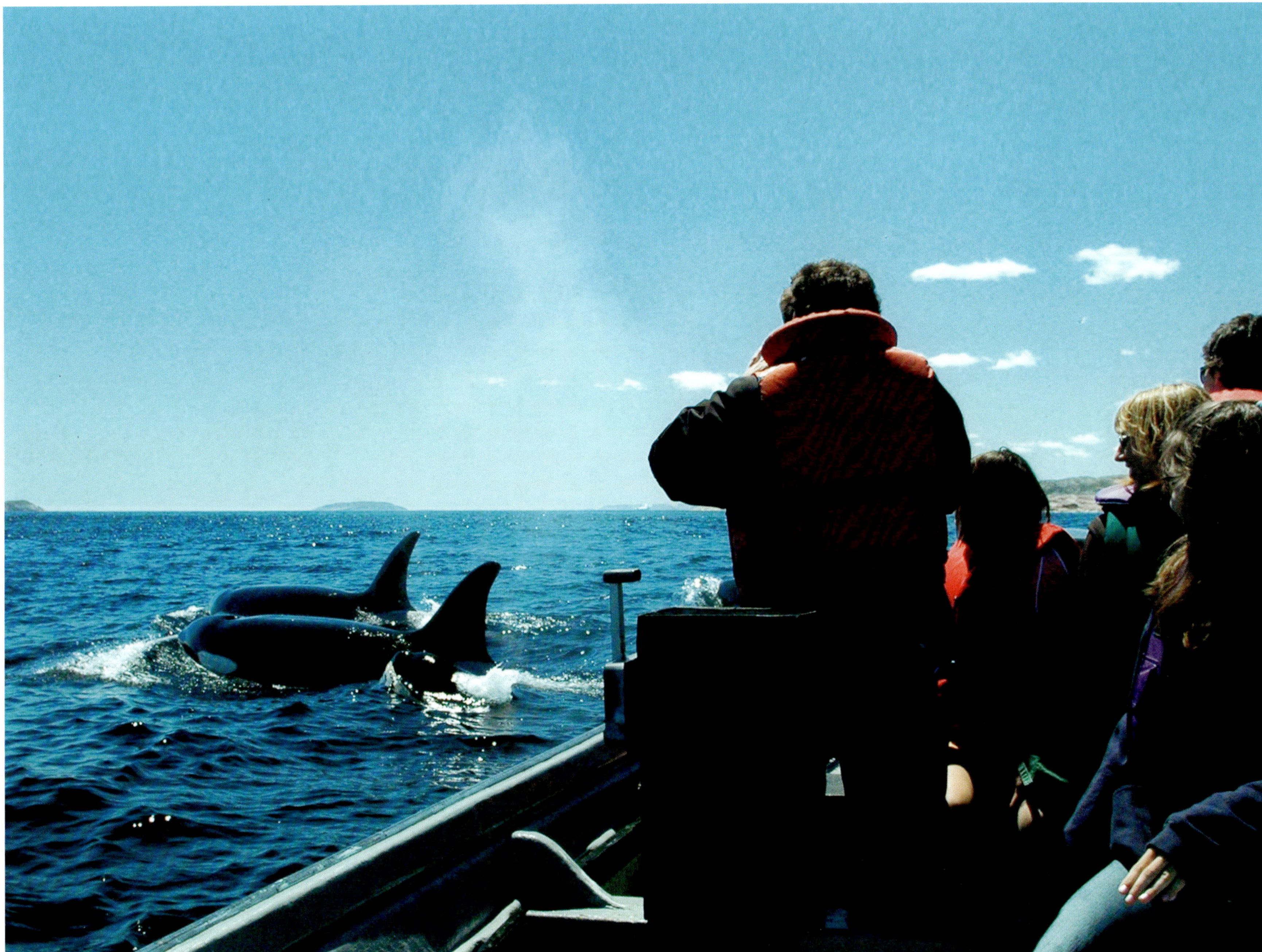

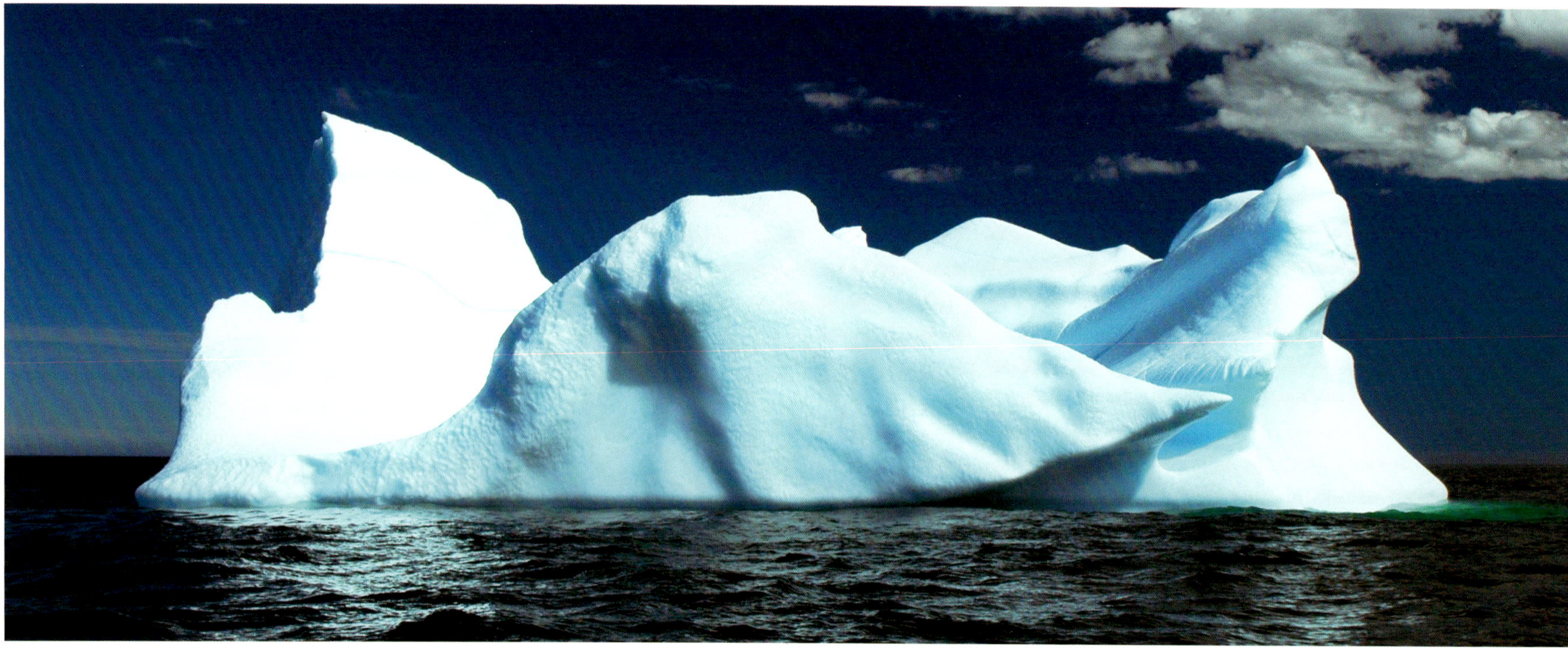

The Iceberg

Kevin

Icebergs are scarce as hens' teeth the latter part of July, even in the best iceberg seasons. From Battle Harbour Island no icebergs were visible. To our surprise, behind the killer whales we sighted an iceberg grounded in the Lodge near Cape Charles, only a ten-minute steam from the whales. On arriving to photograph the iceberg I was told in no uncertain terms that it was going to be zoom lens only for this day. Jacquelyn, still "traumatized" by watching the movie *Titanic,* had had enough "exciting terrifying" experiences for this day.

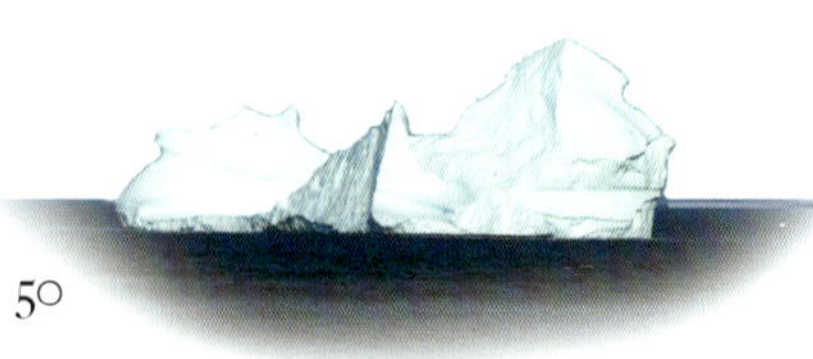

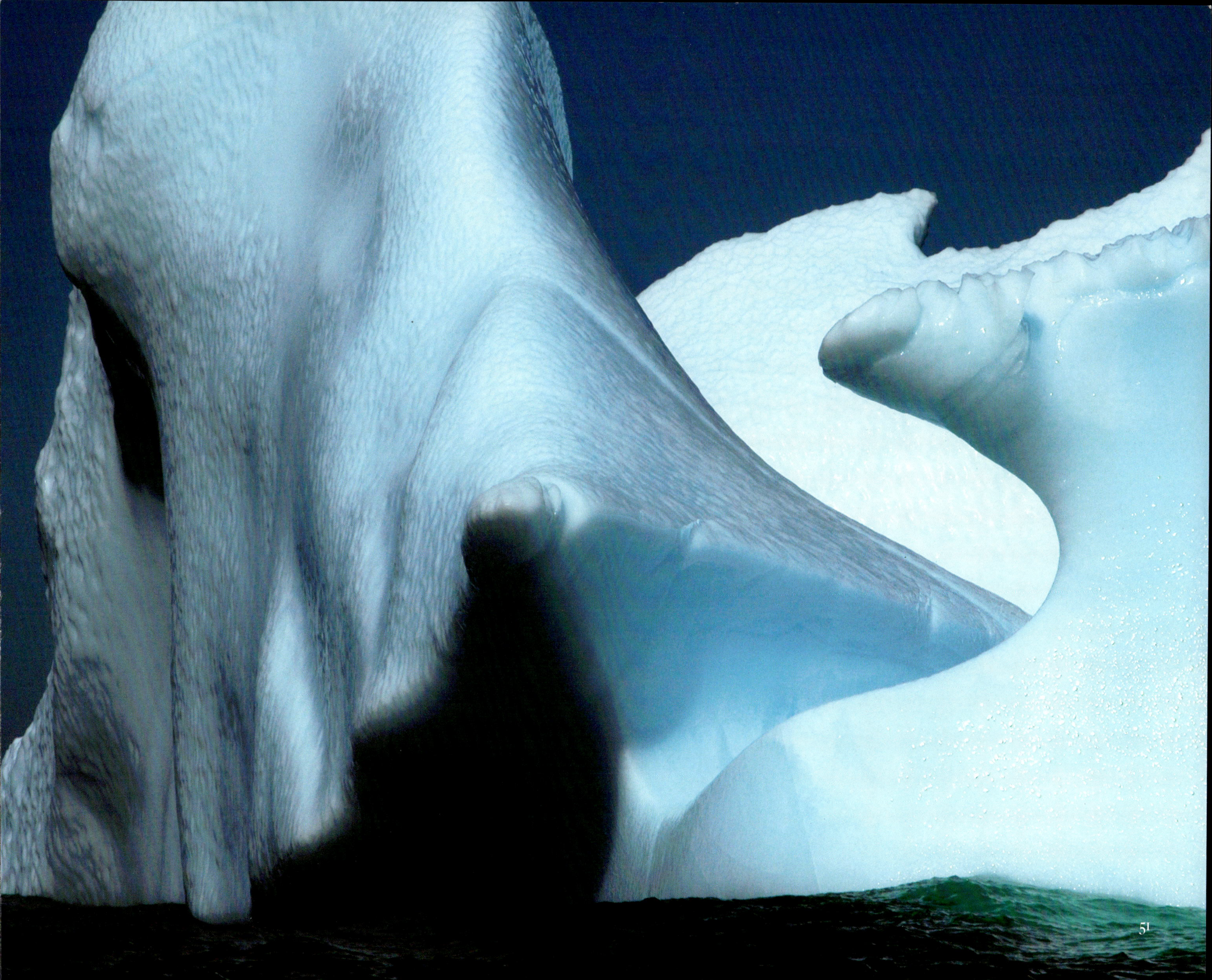

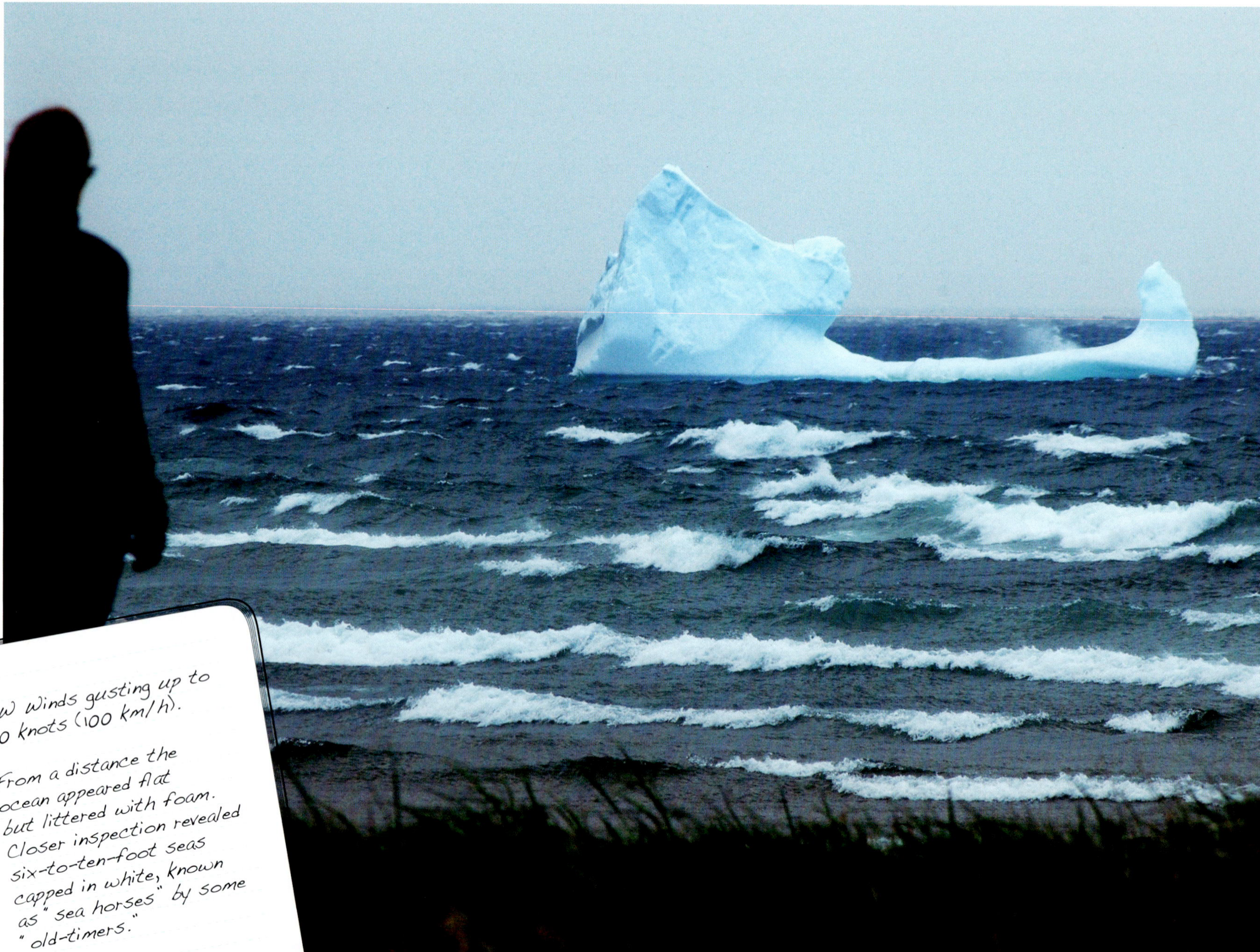

SW winds gusting up to
60 knots (100 km/h).

From a distance the
ocean appeared flat
but littered with foam.
Closer inspection revealed
six-to-ten-foot seas
capped in white, known
as "sea horses" by some
"old-timers."

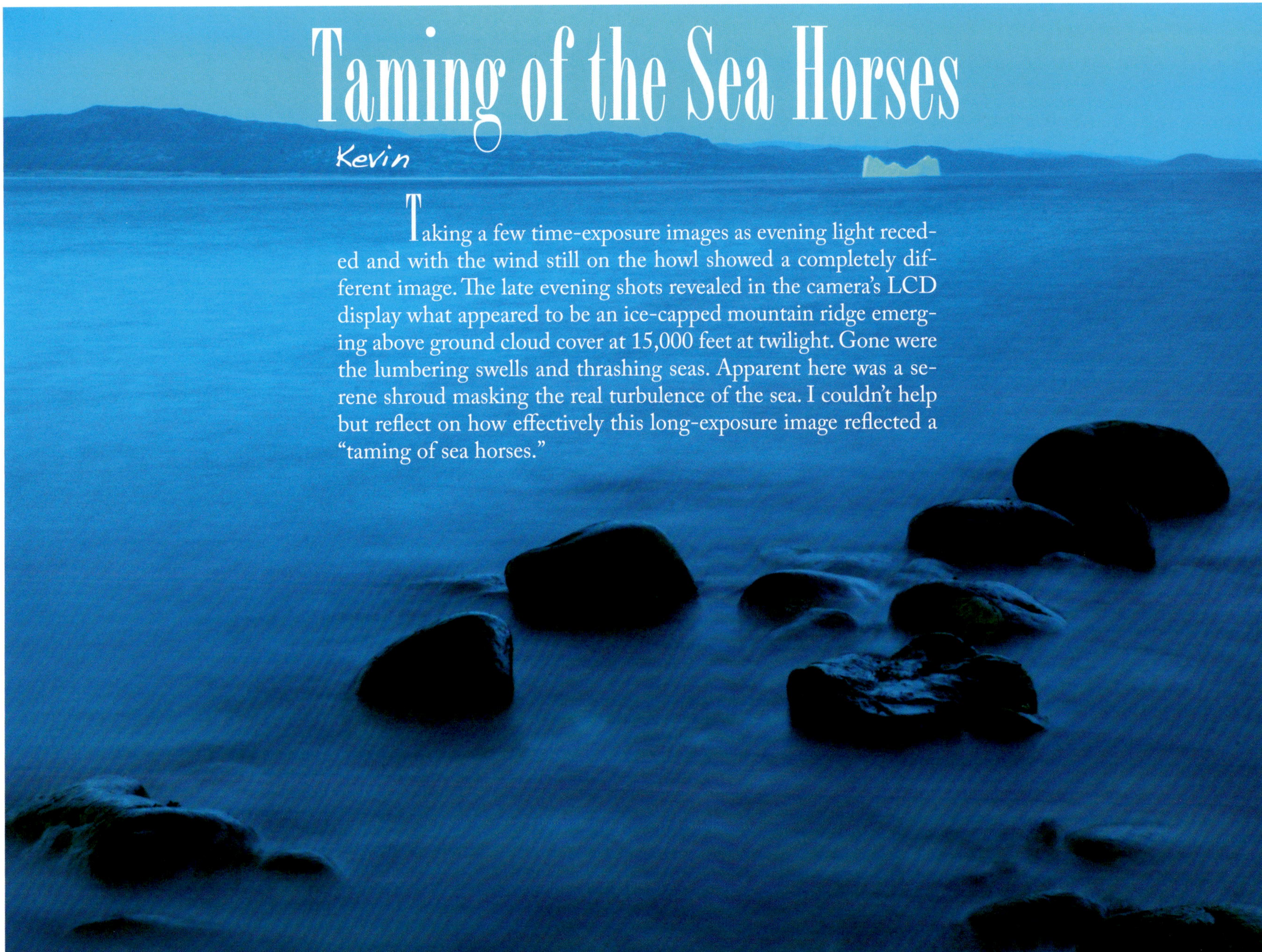

Taming of the Sea Horses

Kevin

Taking a few time-exposure images as evening light receded and with the wind still on the howl showed a completely different image. The late evening shots revealed in the camera's LCD display what appeared to be an ice-capped mountain ridge emerging above ground cloud cover at 15,000 feet at twilight. Gone were the lumbering swells and thrashing seas. Apparent here was a serene shroud masking the real turbulence of the sea. I couldn't help but reflect on how effectively this long-exposure image reflected a "taming of sea horses."

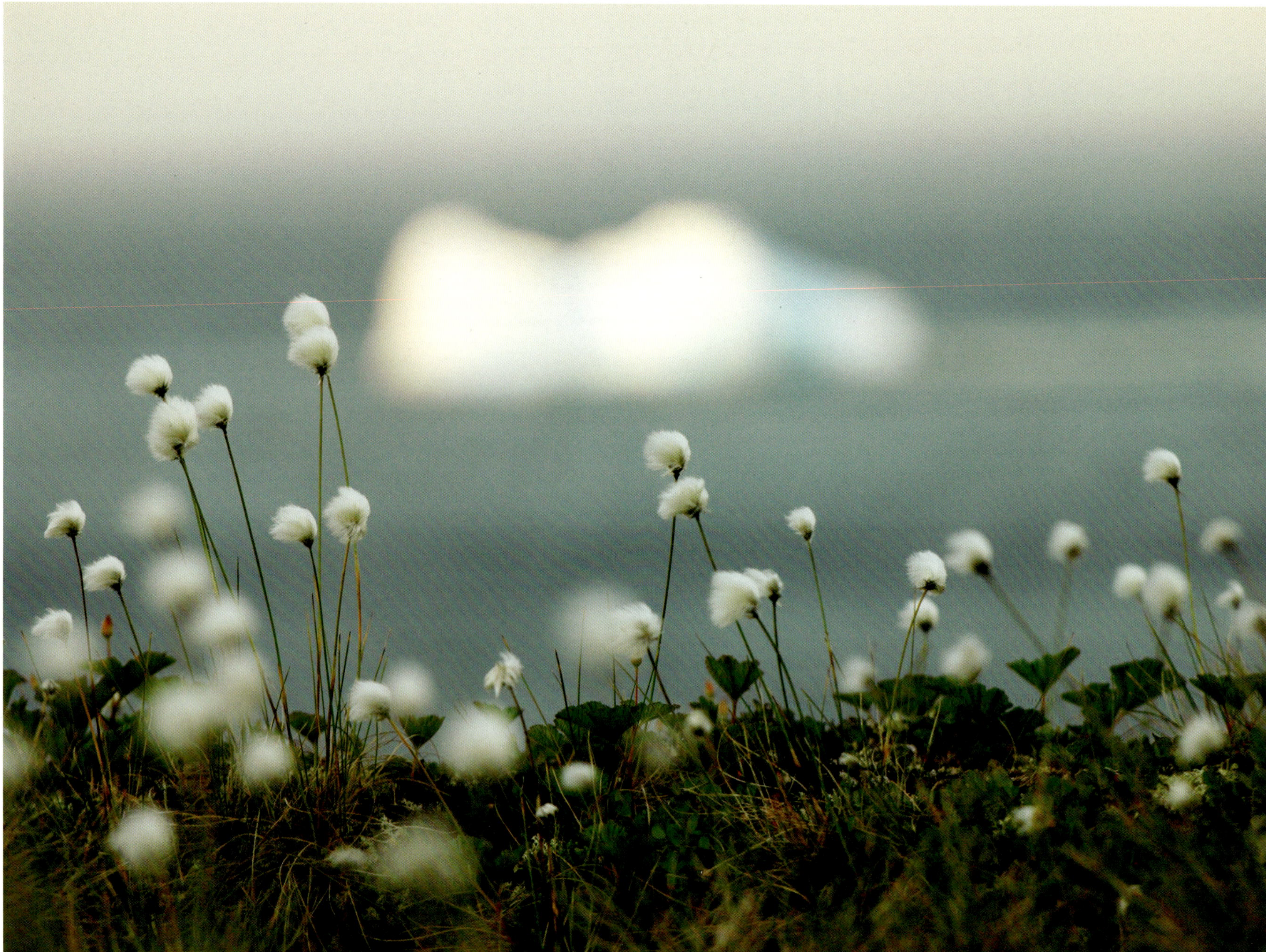

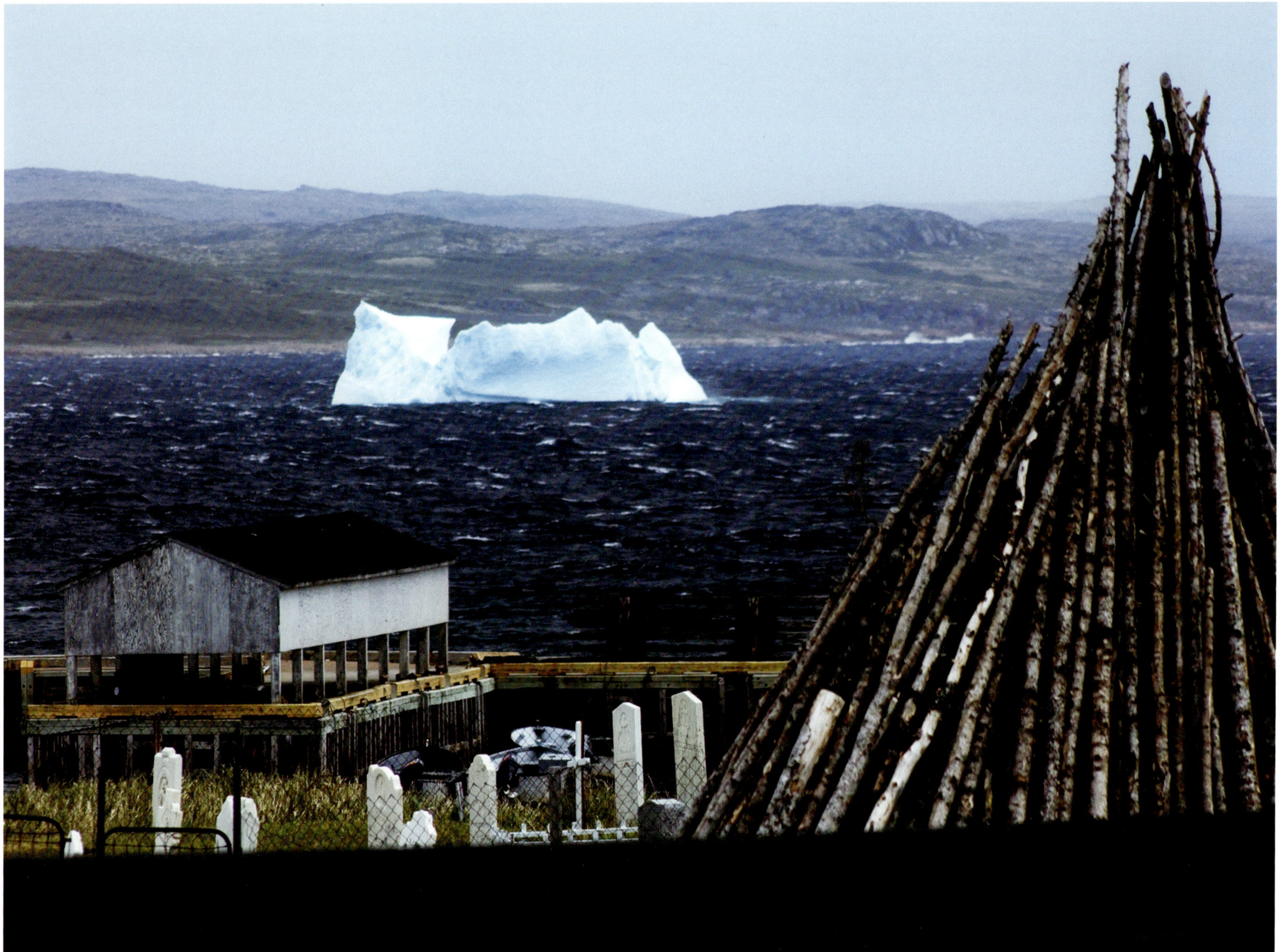

Out of the Fog

Kevin

Next question: take camera gear or leave it home? Weatherwise, the Strait of Belle Isle is a dynamic place. Dense fog moves in and out depending on a wide variety of wind patterns and their interaction with the warm land mass in summer and cool waters descending from the Labrador Sea that swish up and down the strait with speeds up to eight knots. Hence it was a resounding yes to taking the camera gear; just never know what I might come across.

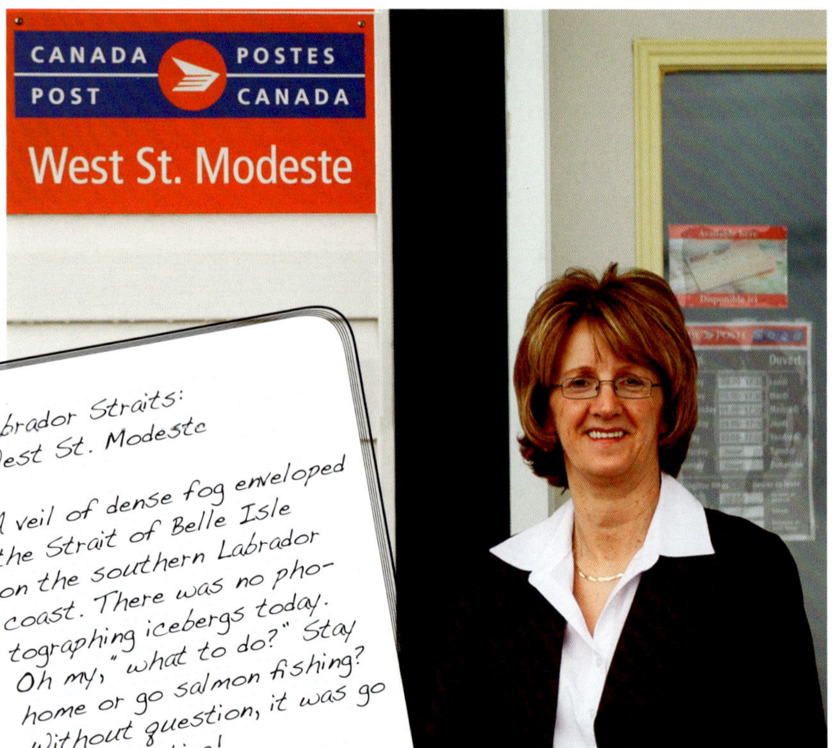

Labrador Straits:
West St. Modeste

A veil of dense fog enveloped the Strait of Belle Isle on the southern Labrador coast. There was no photographing icebergs today. Oh my, "what to do?" Stay home or go salmon fishing? Without question, it was go salmon fishing!

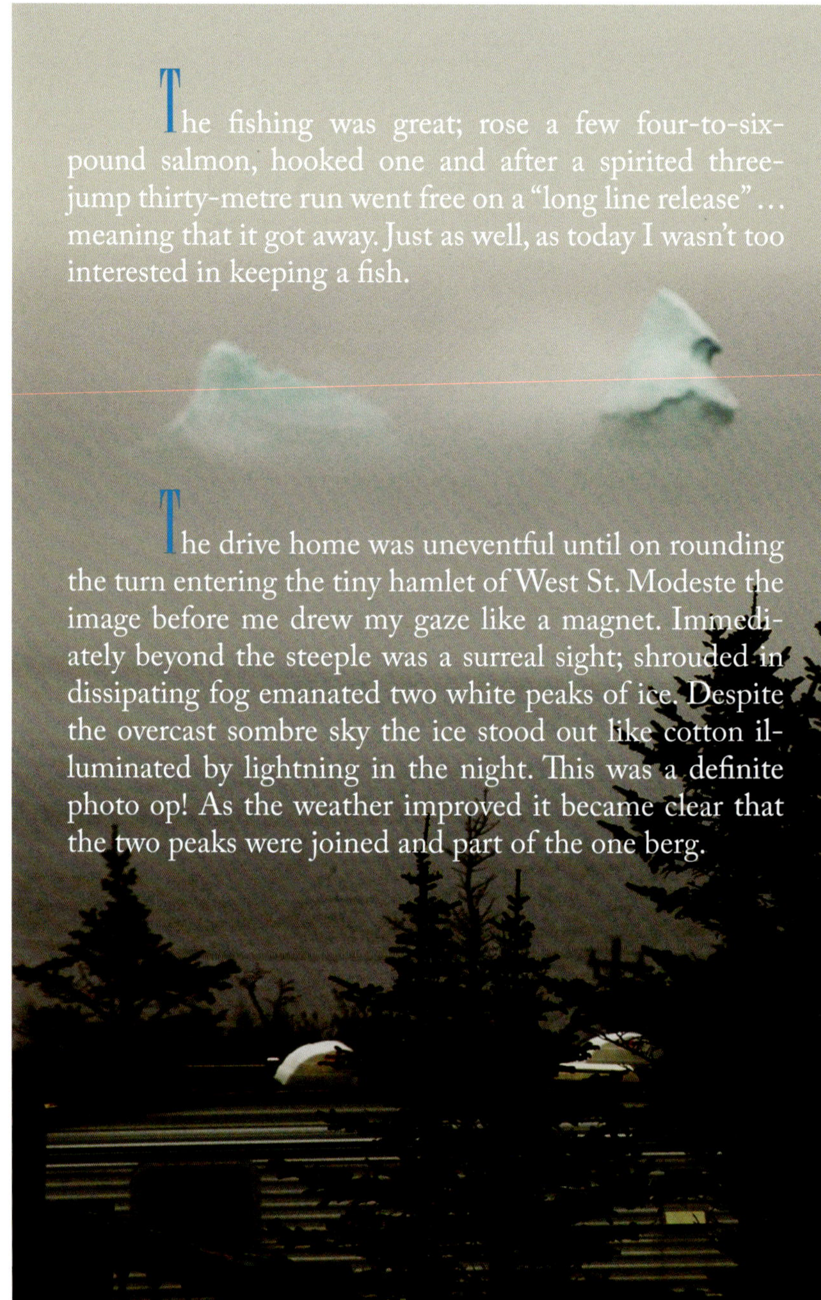

The fishing was great; rose a few four-to-six-pound salmon, hooked one and after a spirited three-jump thirty-metre run went free on a "long line release"... meaning that it got away. Just as well, as today I wasn't too interested in keeping a fish.

The drive home was uneventful until on rounding the turn entering the tiny hamlet of West St. Modeste the image before me drew my gaze like a magnet. Immediately beyond the steeple was a surreal sight; shrouded in dissipating fog emanated two white peaks of ice. Despite the overcast sombre sky the ice stood out like cotton illuminated by lightning in the night. This was a definite photo op! As the weather improved it became clear that the two peaks were joined and part of the one berg.

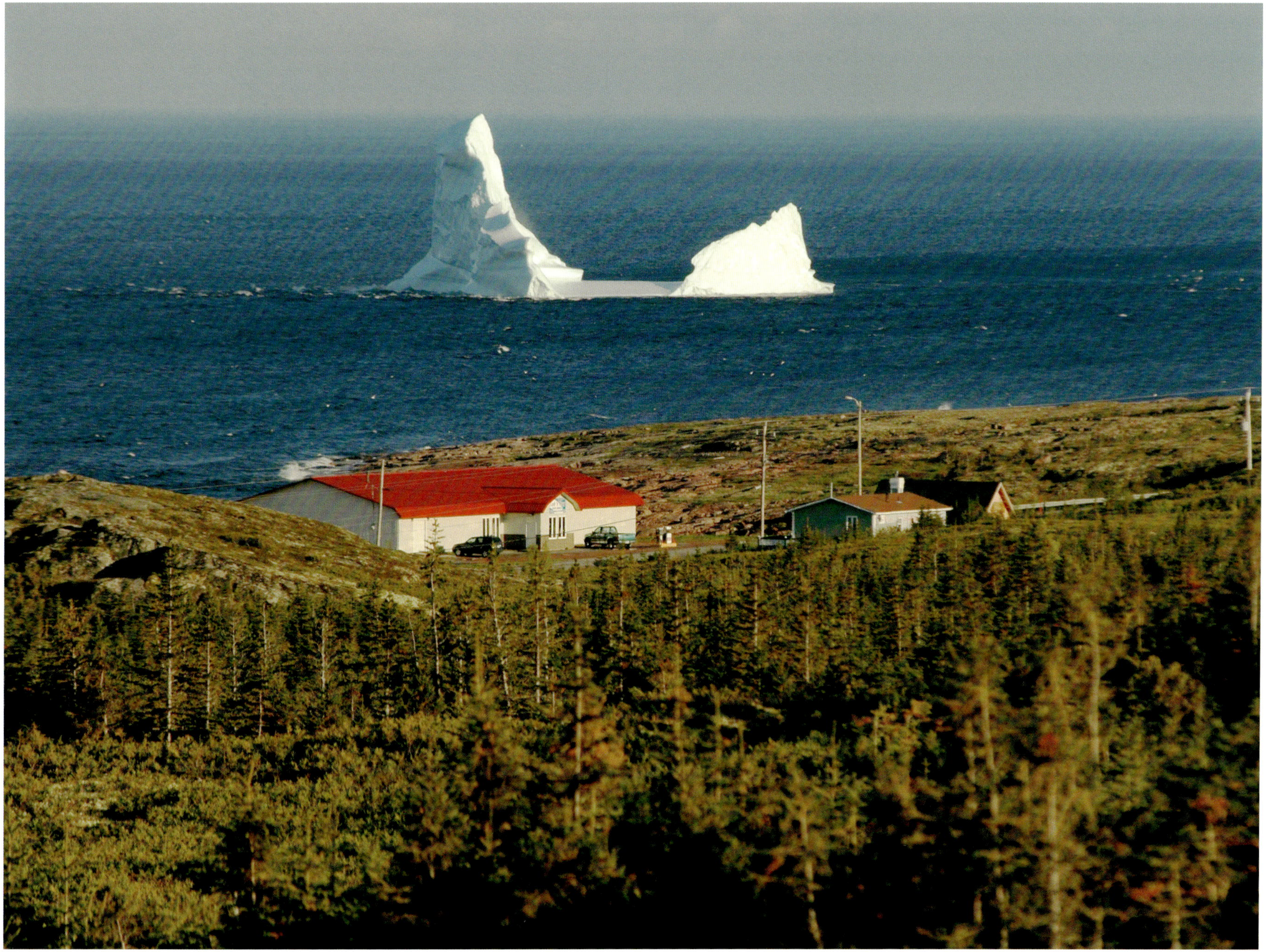

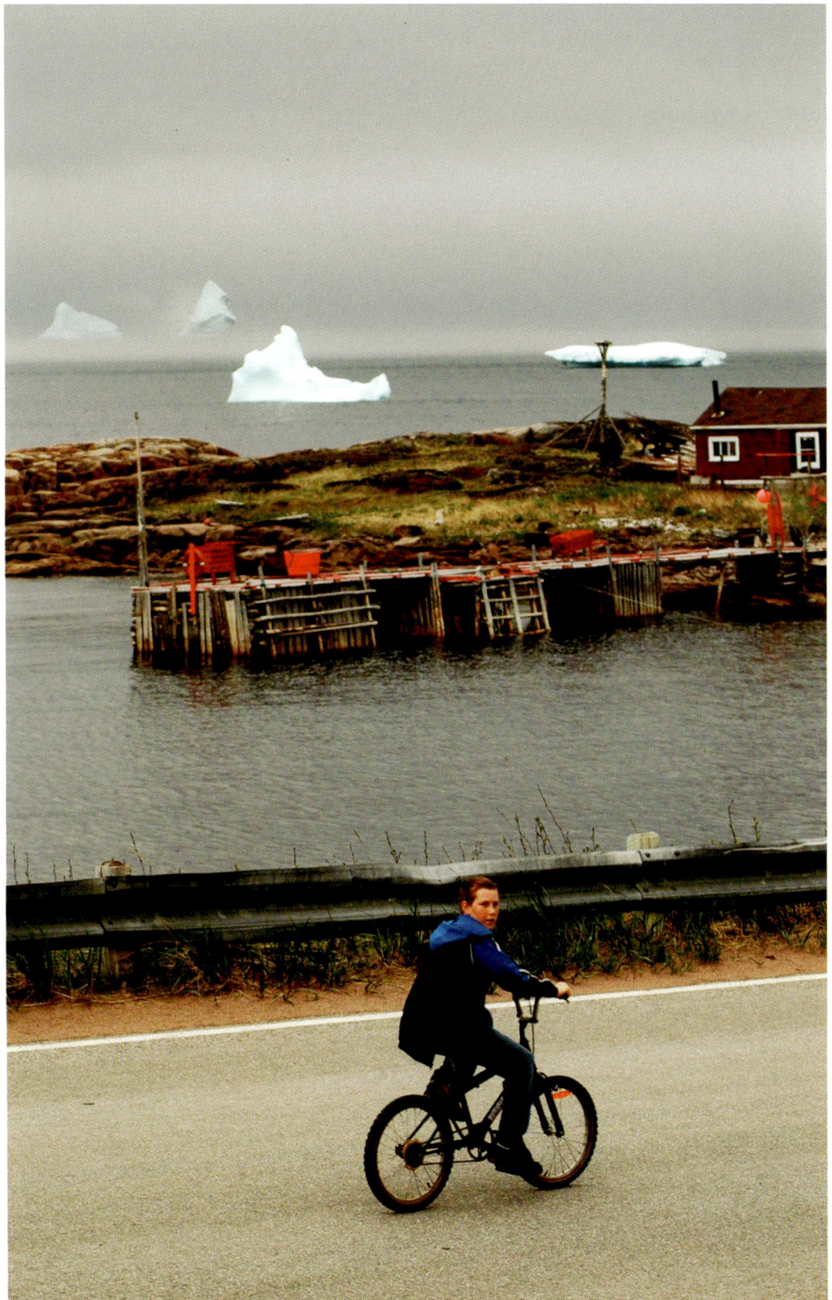

Sewing Machine Knees

Kevin

With butterflies in my belly and with "sewing machine knees" I stood behind the tripod-mounted camera. This was something never to be seen again!

Three layers comprised the image: anchoring the image was a traditional wharf and traditional "store" (local name for shed); the middle layer was two grounded but spaced icebergs; and the outside layer was a larger iceberg moving southwest "down" the Straits.

The larger free-floating iceberg lumbered along like a giant steamship. I knew it wouldn't be long before this larger berg moved in position for the ideal shot . . . . placed perfectly between the two grounded bergs, making symmetry appear natural, not contrived.

It was ironic that while I waited, Stephen O'Dell riding his bike down the road in front of me was more curious of me than the events unfolding behind him. To him it was a normal anomaly but to me it was ethereal and I couldn't help but think: "How lucky he is!"

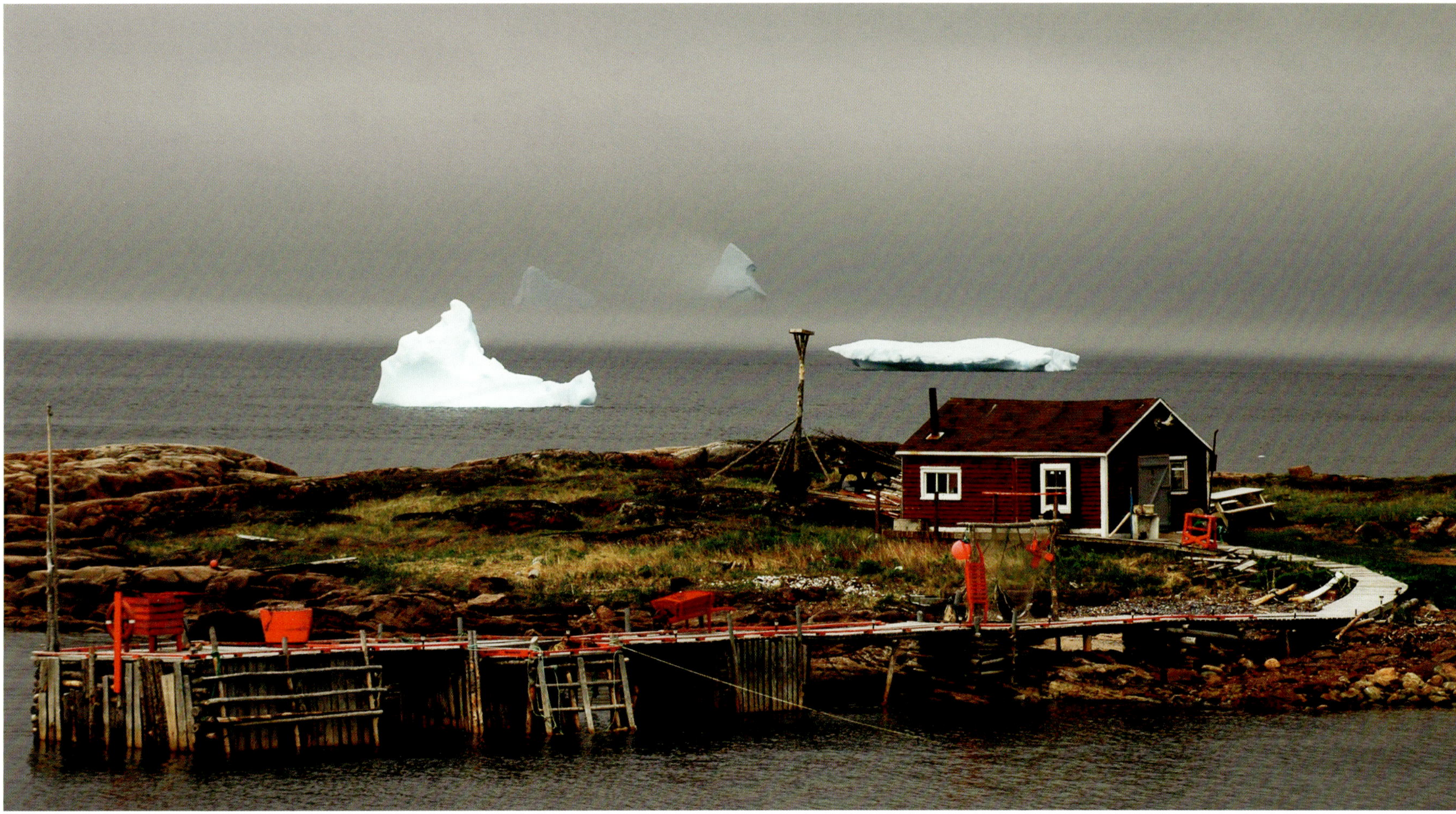

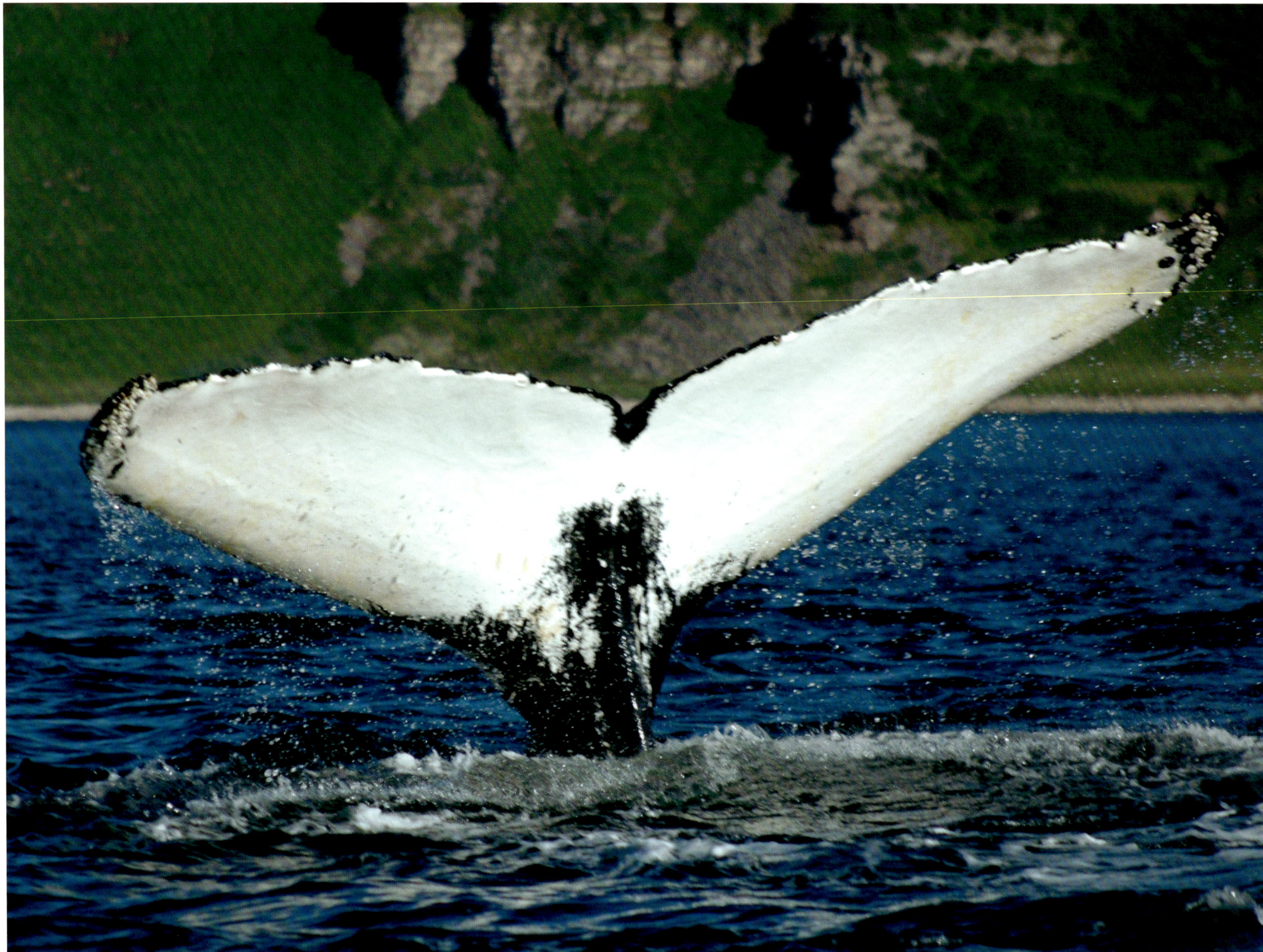

Whales like Sailboats

Kevin

This morning, rising with the sun, shouldering my camera gear, I meandered to the lookout atop Eastern Hill. From this lookout the previous evening, all that was visible was a "black of fog." Hence it was not surprising that my first scan of the morning's horizon revealed a single column of foggy mist. Subsequent movement at the water's surface below the fog offered a clear explanation for the morning foggy plume of mist. It was a whale! What kind of whale? A broad tail displaying the whale's fingerprint revealed its gene pool; it was a humpback! A more detailed scan of the ocean in front of me showed another couple of dozen spouting humpback whales.

Not only were there whales but a beautiful piece of Greenland ice, caressed and sculpted by the ocean water, air, and sun. A midsummer's dream; I had to get out in a boat! But how? Everyone was asleep! Rushing home I called and woke my brother-in-law, Joe Fowler, to captain his boat. Groggily, he agreed, thank God!

There were lots of whales and obviously lots of bait fish at the surface, for the whales only sparingly made the terminal dive to show their tails. This made it difficult getting the signature shot I was looking for . . . the whale tail and the iceberg . . . . Patience proved its virtue, as we were fortunate to catch a few whales willing to dive in front of the iceberg making this our newest "best day with whales ever!"

Labrador Straits:
Capstan Island

Some years ago I overheard Francis Smith's succinct description of a large pod of humpback whales in the Labrador Straits: ". . . whales like sailboats!" It turned out to be our best day with whales ever!

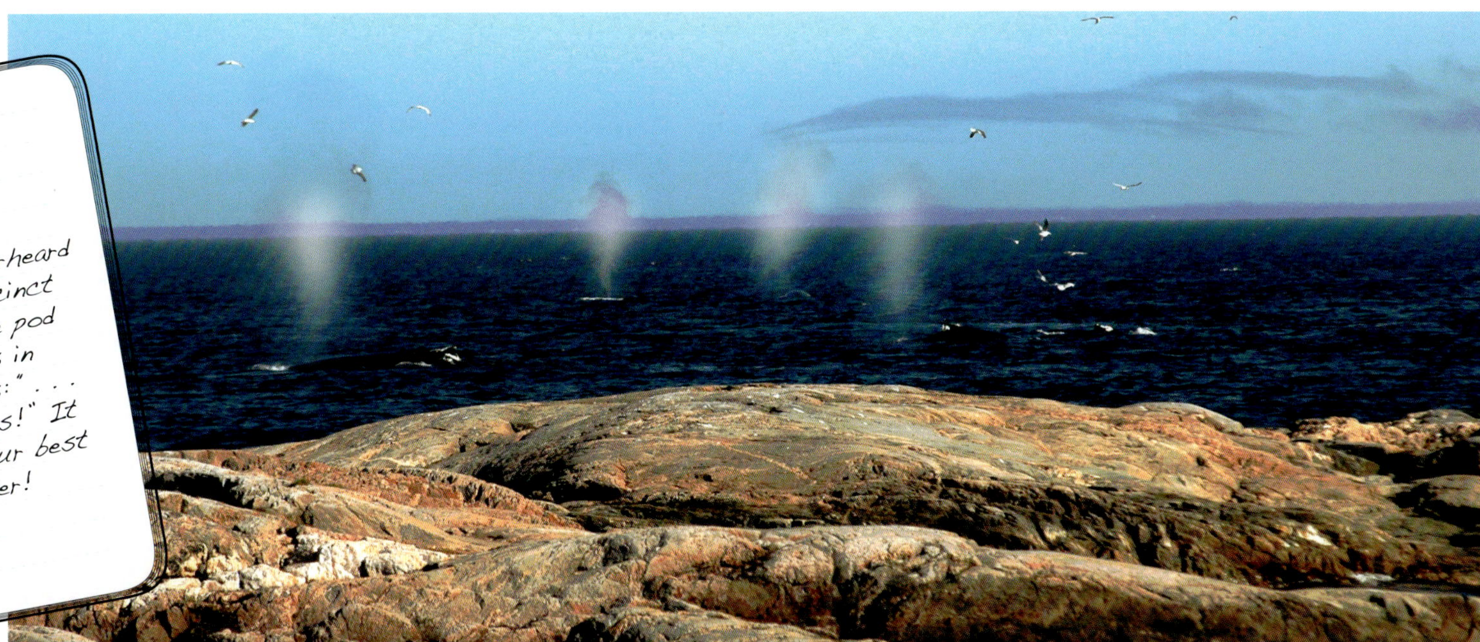

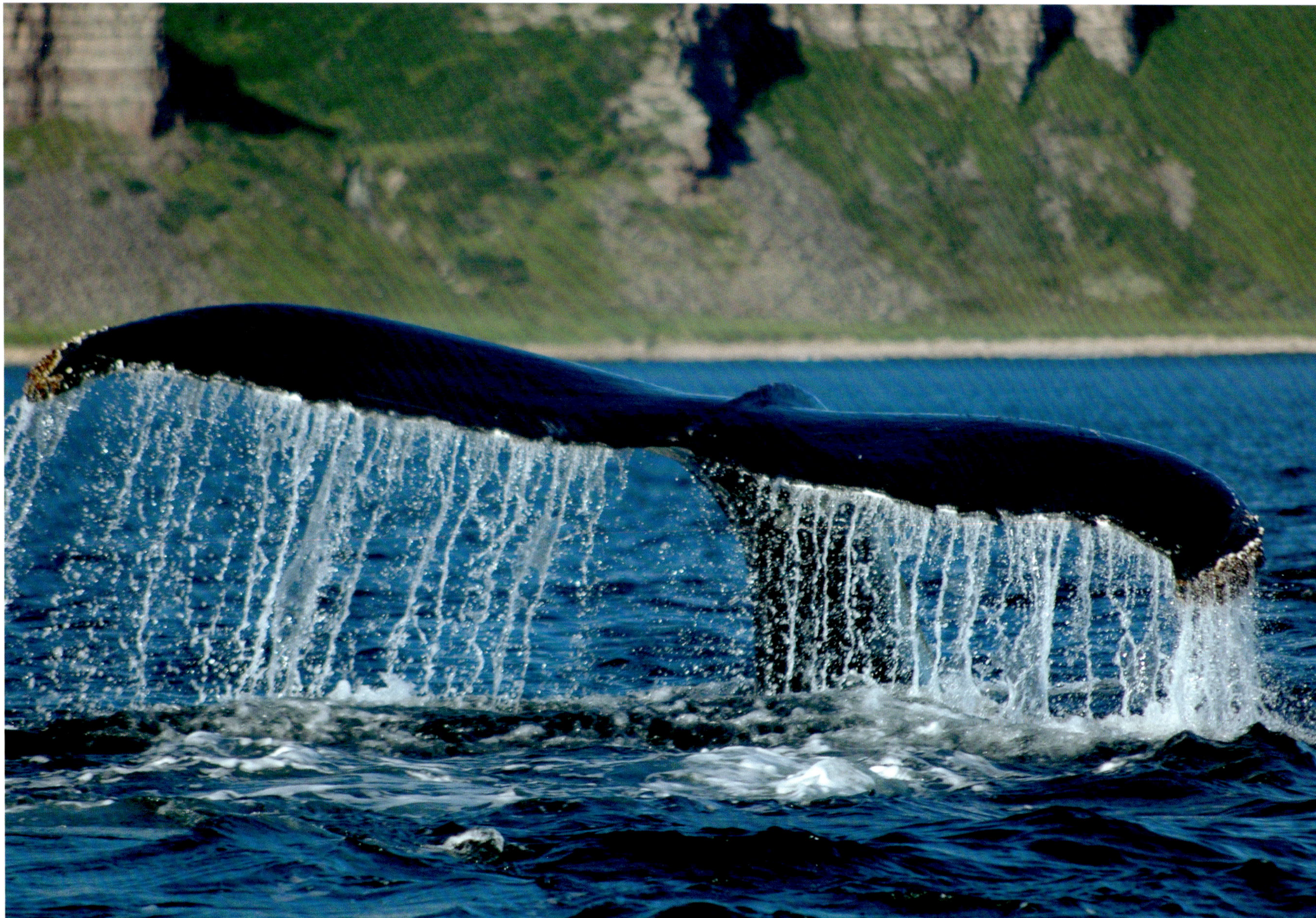

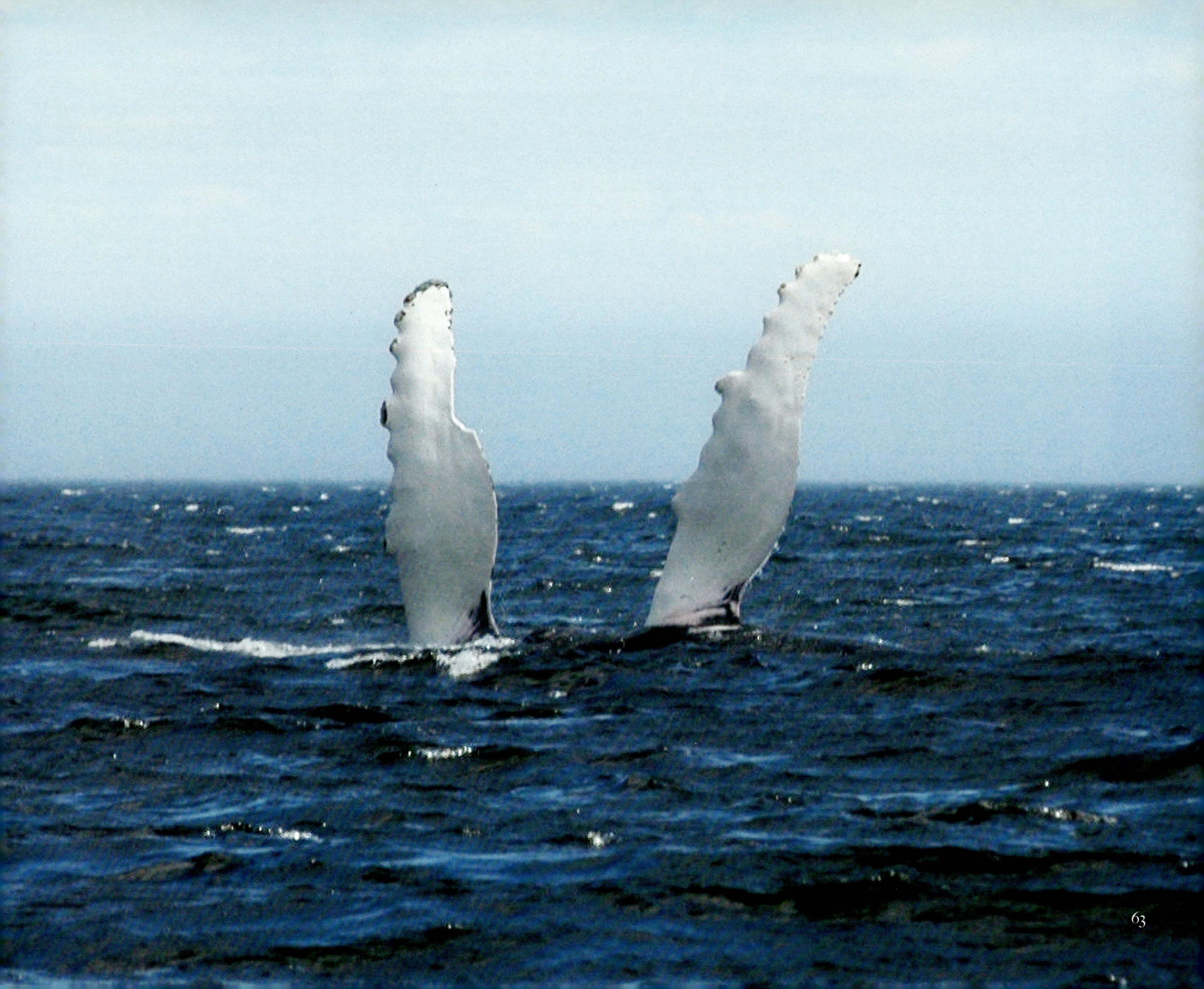

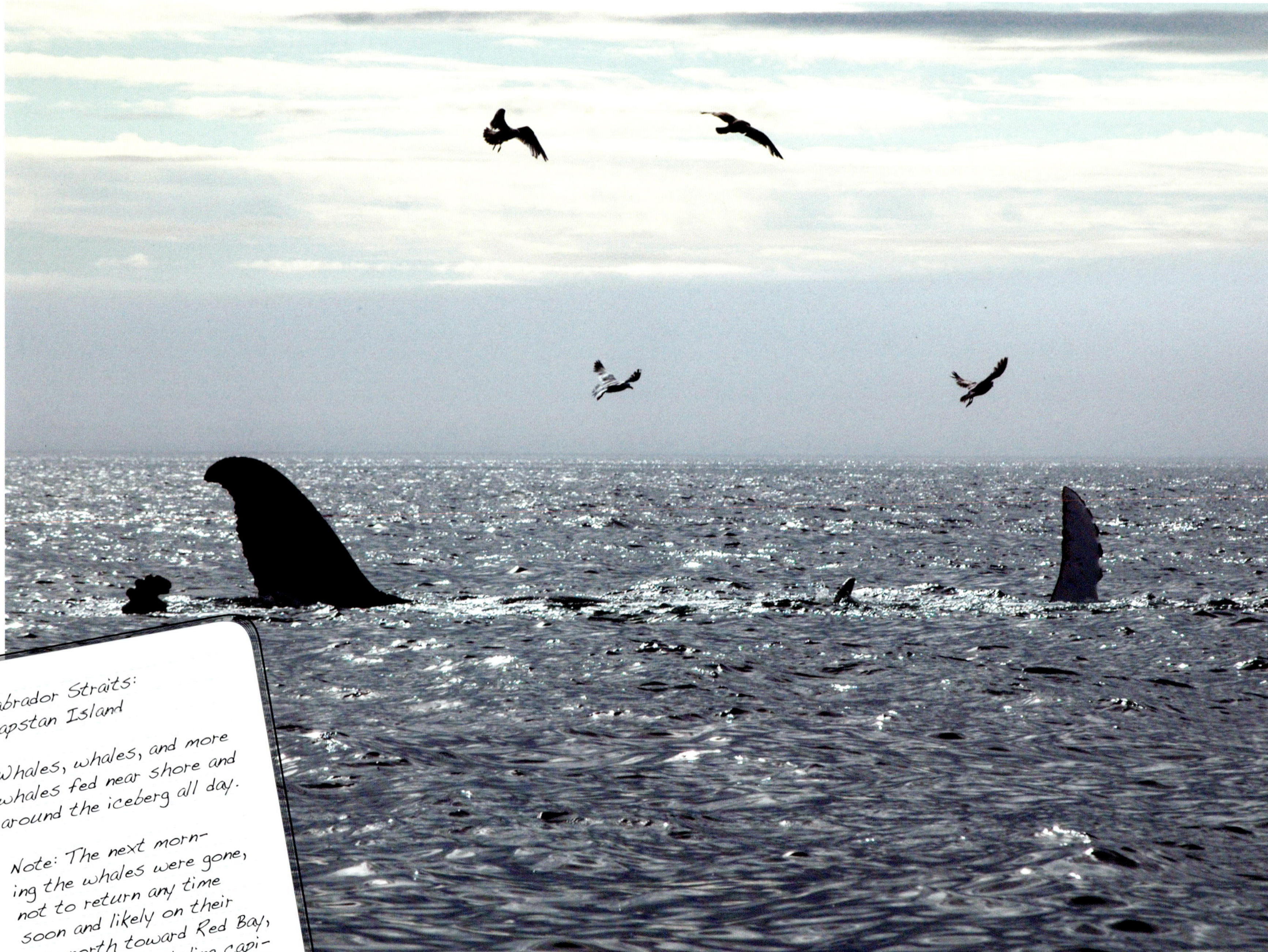

Labrador Straits:
Capstan Island

Whales, whales, and more whales fed near shore and around the iceberg all day.

Note: The next morning the whales were gone, not to return any time soon and likely on their way north toward Red Bay, 16th-century whaling capital of the world.

Supper by the Sea

Kevin

After watching the whales since daylight, by late afternoon I began to notice some predictable patterns. A couple of whales appeared to be feeding together all day near the outer southeast reaches of Capstan Island. It was so close to land I excitedly took my supper and headed for that point of land, getting me as close to the whales and iceberg as possible, without being in a boat.

There, two humpbacks, side by side just metres from shore, repeatedly swam a similar route. Beginning in the cove, they followed the shoreline swimming on or near the surface toward the open water of the Straits. Once they were a short distance past the island, the whales disappeared for a couple of minutes before repeating the same outward journey. I could only surmise that when the whales disappeared they swam in the cove deep underwater, feeding and/or driving the fish to the surface.

While I watched this spectacle a large powerboat sidled up to the iceberg. The boat made the iceberg appear smaller than what I had seen up close earlier in the day. Then the powerboat lowered its lifeboat which was miniscule compared to the powerboat and the iceberg. While the large boat stayed in close proximity to the iceberg, the lifeboat went *zoom, zoom* frolicking for and with whales.

It was not until later that evening that I found out that it wasn't a "normal" powerboat; it was a yacht, eighty feet long and three storeys high, visiting from Florida. Further to the story was that one of the passengers had stood on the iceberg. On hearing these reports I felt lucky not to have witnessed a catastrophe.

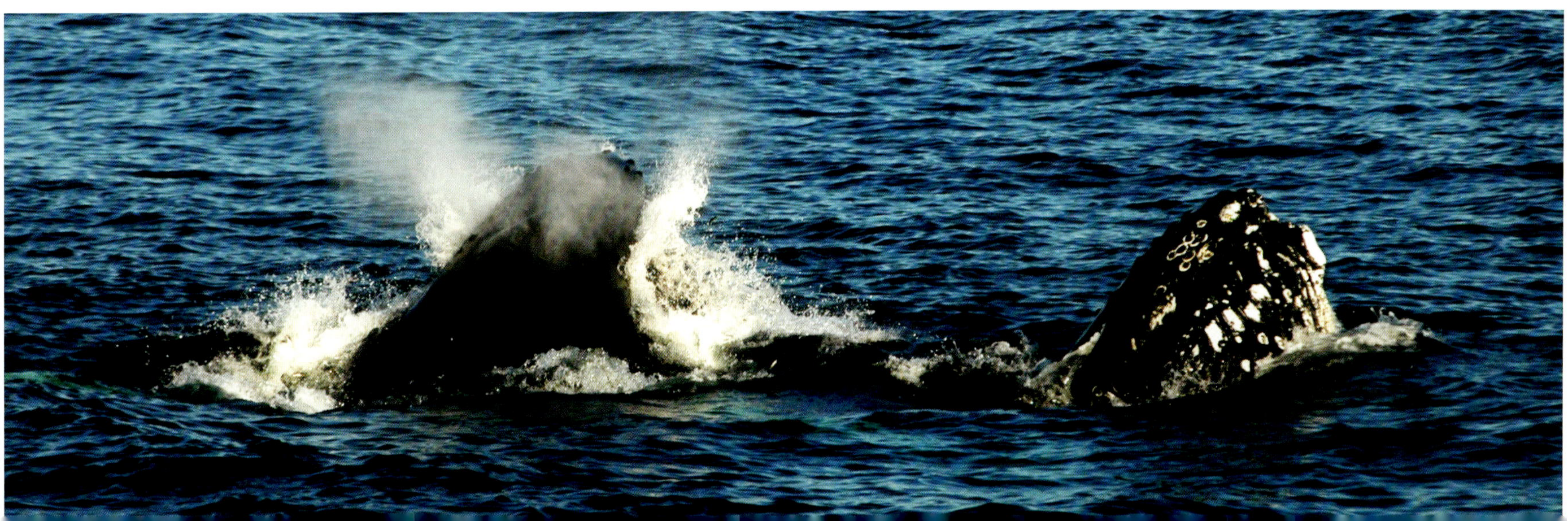

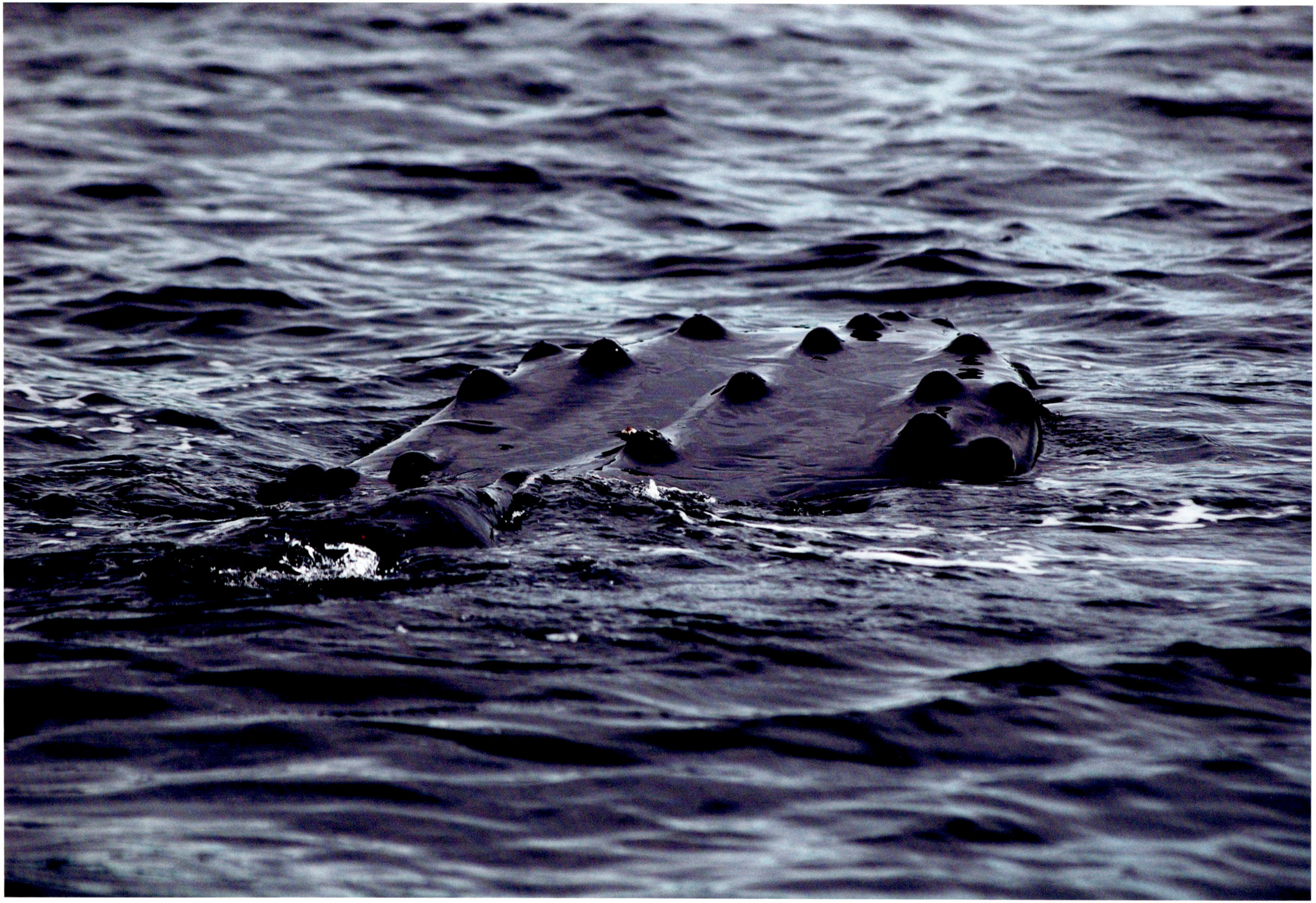

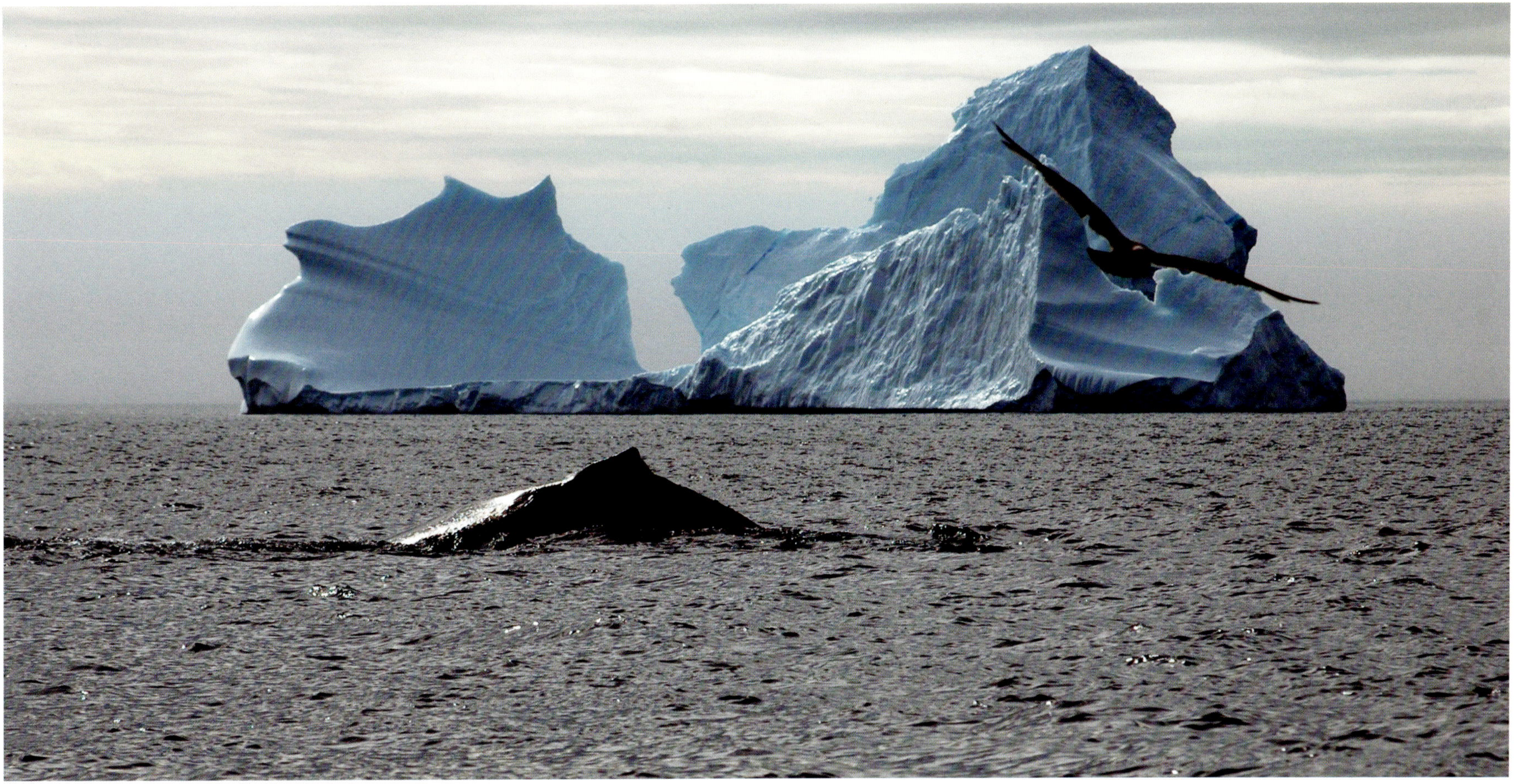

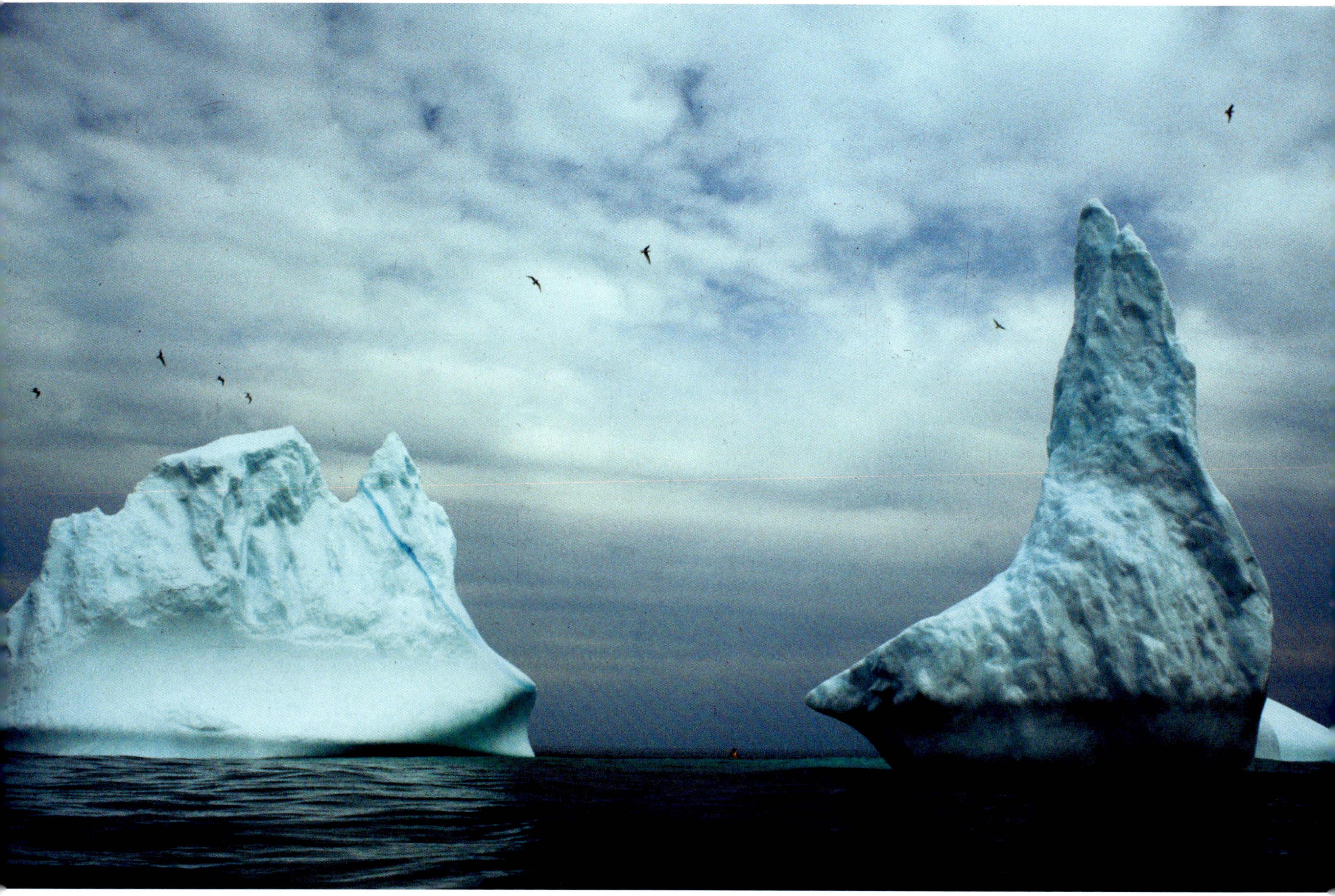

The Twin Towers

Kevin

Paddling with me today was brother-in-law David, his last job being an engineer with NASA. David's adaptable personality endears him to local cultures wherever he goes. Born in Ohio, and widely travelled, he is up for whatever the local culture has to offer. Today was no exception and by his account was "an ethereal experience!"

A short, ten-minute paddle put us within the iceberg's "high-risk buffer zone," meaning we were fine as long as the berg didn't split, founder or roll. If the iceberg did become active, we would likely be involved. Sea state at the time was slow rolling swells: turbulent on shore, civil and calm in open sea.

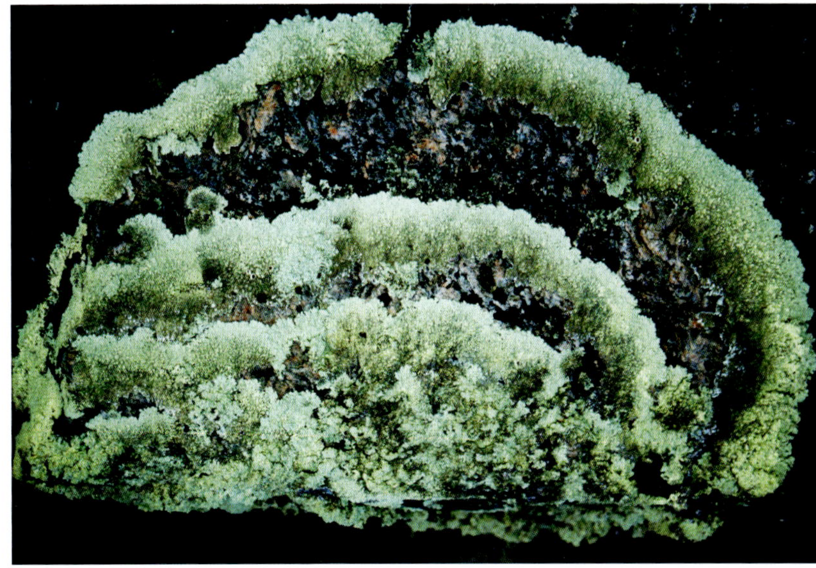

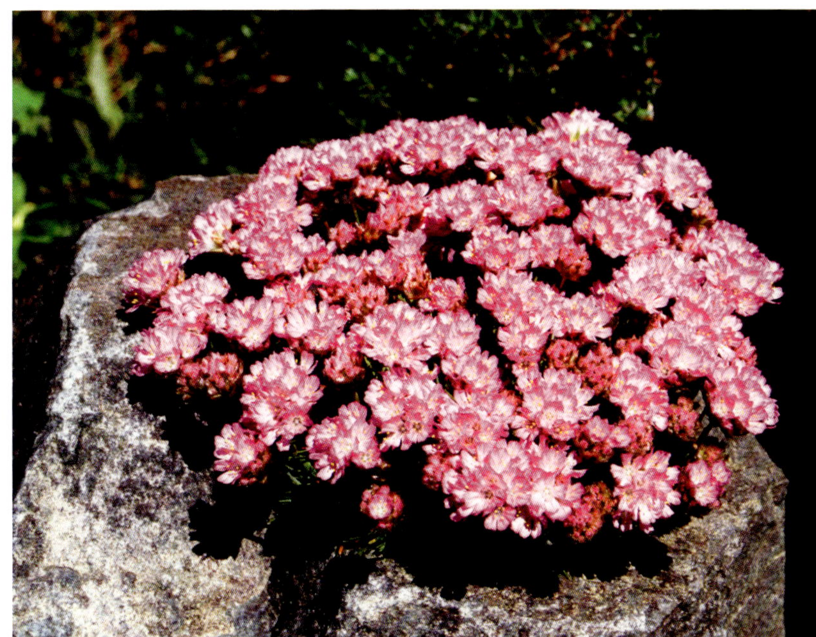

Labrador Straits
L'Anse au Clair
Journal: July 8, 1999

The iceberg at the mouth of L'Anse au Clair bay stood in the distance like two towers. As I paddled away from shore, the last words from my five-year-old son Thomas (spoken like his mother) were: "Dad, don't go between them!"

Attractive risk characterizes the "fatal-like attraction" of people in boats to icebergs. Curiosity lured us closer. Adventure lured us between the towers! A sea swell converging between the two towers rose and fell like an elevator in slow motion. Paddling slowly above the bridge of ice connecting the two towers of ice, a sense of ambivalence embraced me. Awestruck euphoria at being here, countered by my son's last words: "Dad, don't paddle between the towers!" What were we doing? Both of us had young children. The iceberg had been grounded here for a week . . . it was a ticking time bomb! Next time, don't paddle between towers!

Our paddle today featured another highlight: spending a half-hour with humpback whales off Forteau Bay. At one point a whale, three times the length of my seventeen-foot kayak, swam just metres below the hull of my boat.

Today, luck was on our side . . . whales and icebergs. Within twenty-four hours, the iceberg was gone, having foundered and pushed southeast by the strong current renowned in the Strait of Belle Isle.

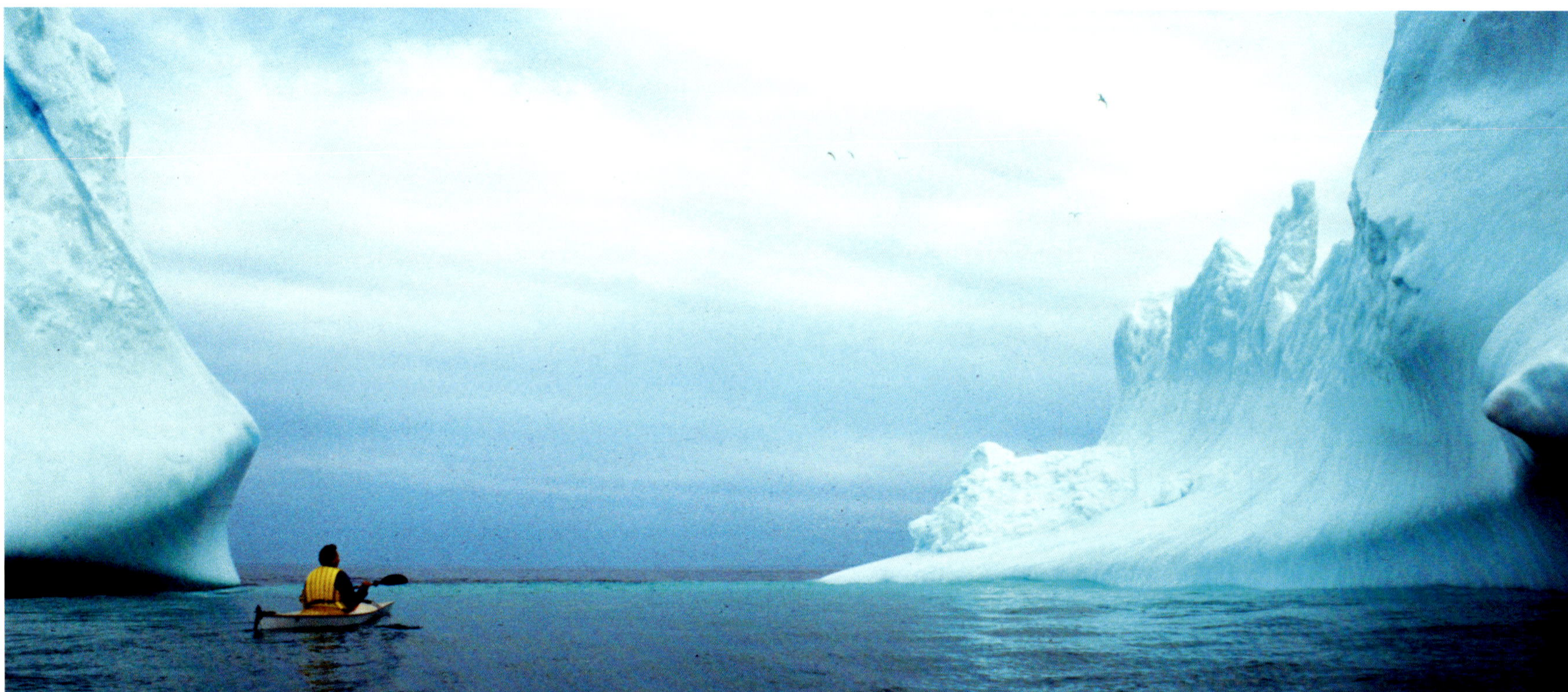

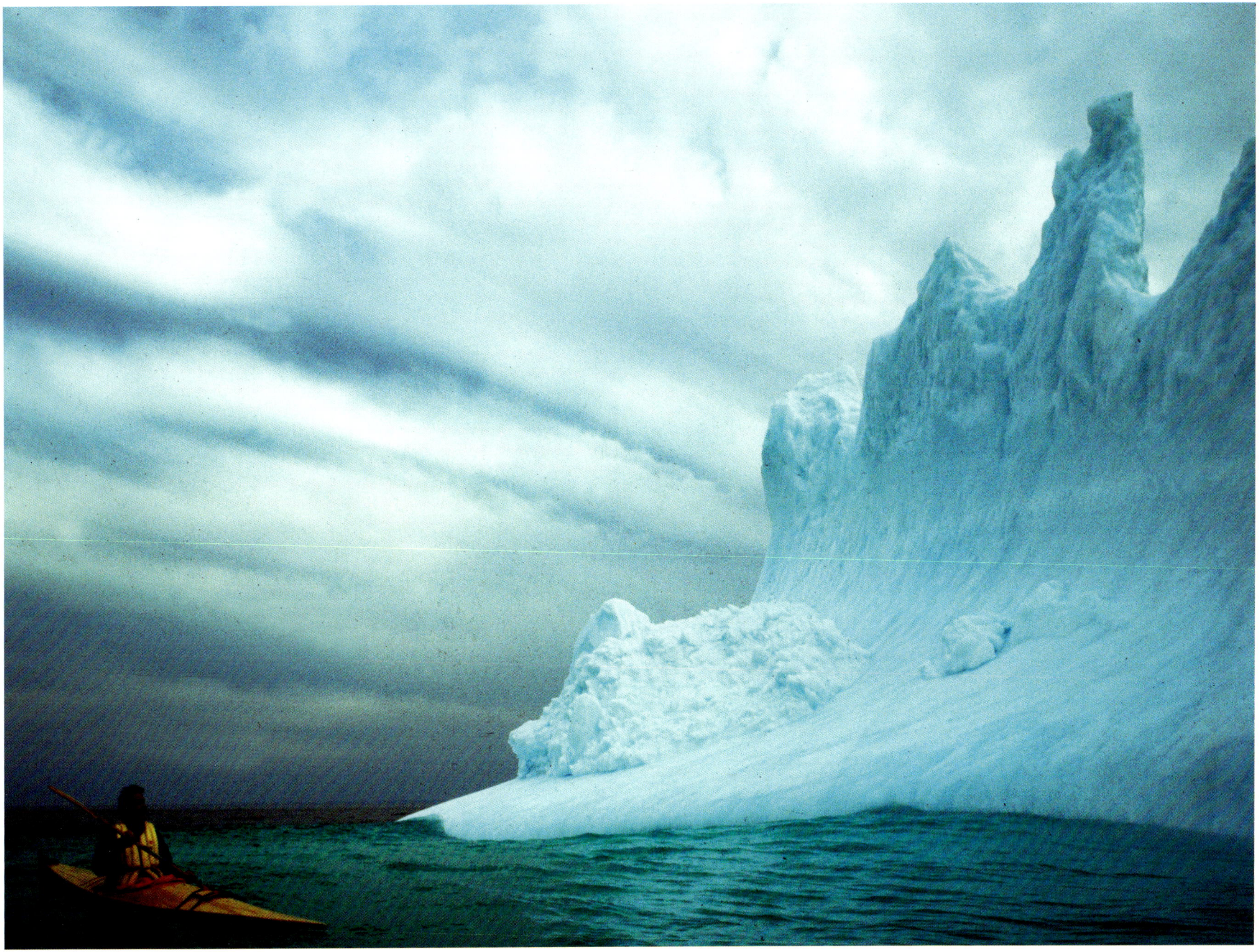

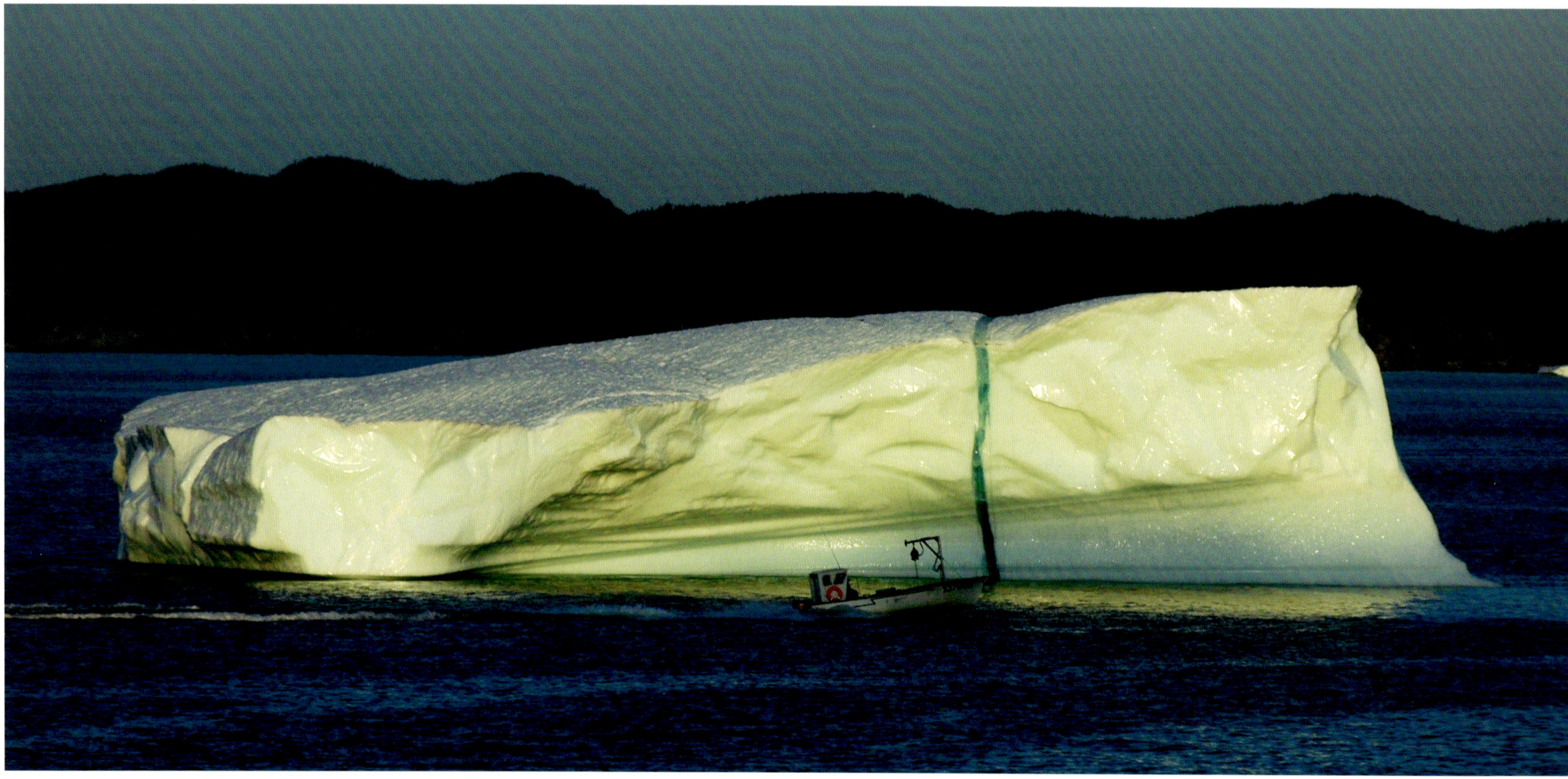

Newfoundland

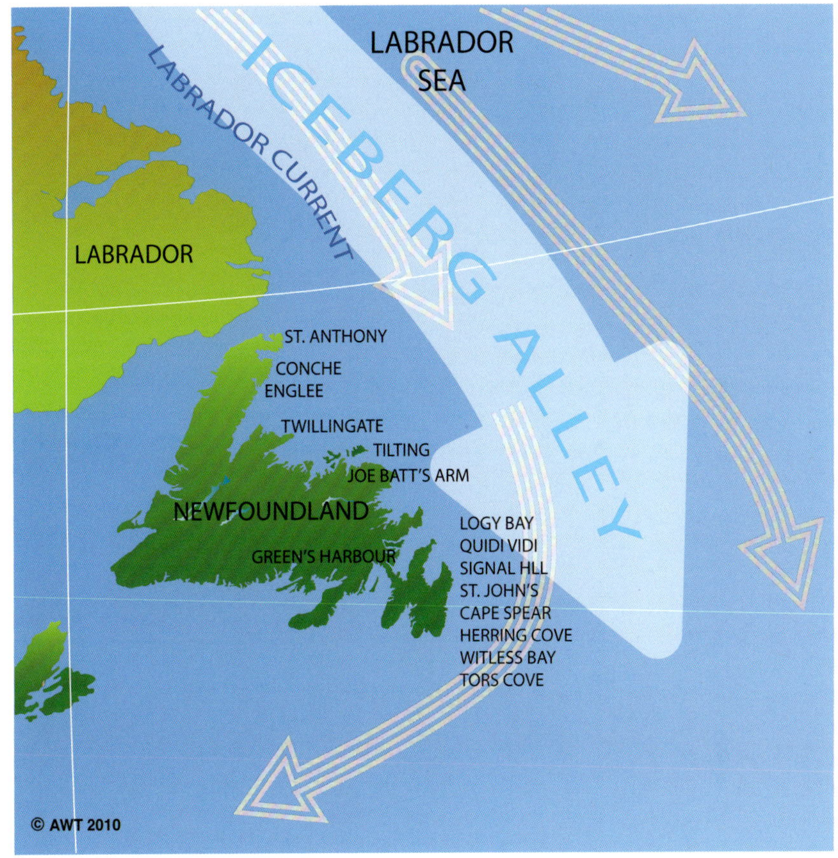

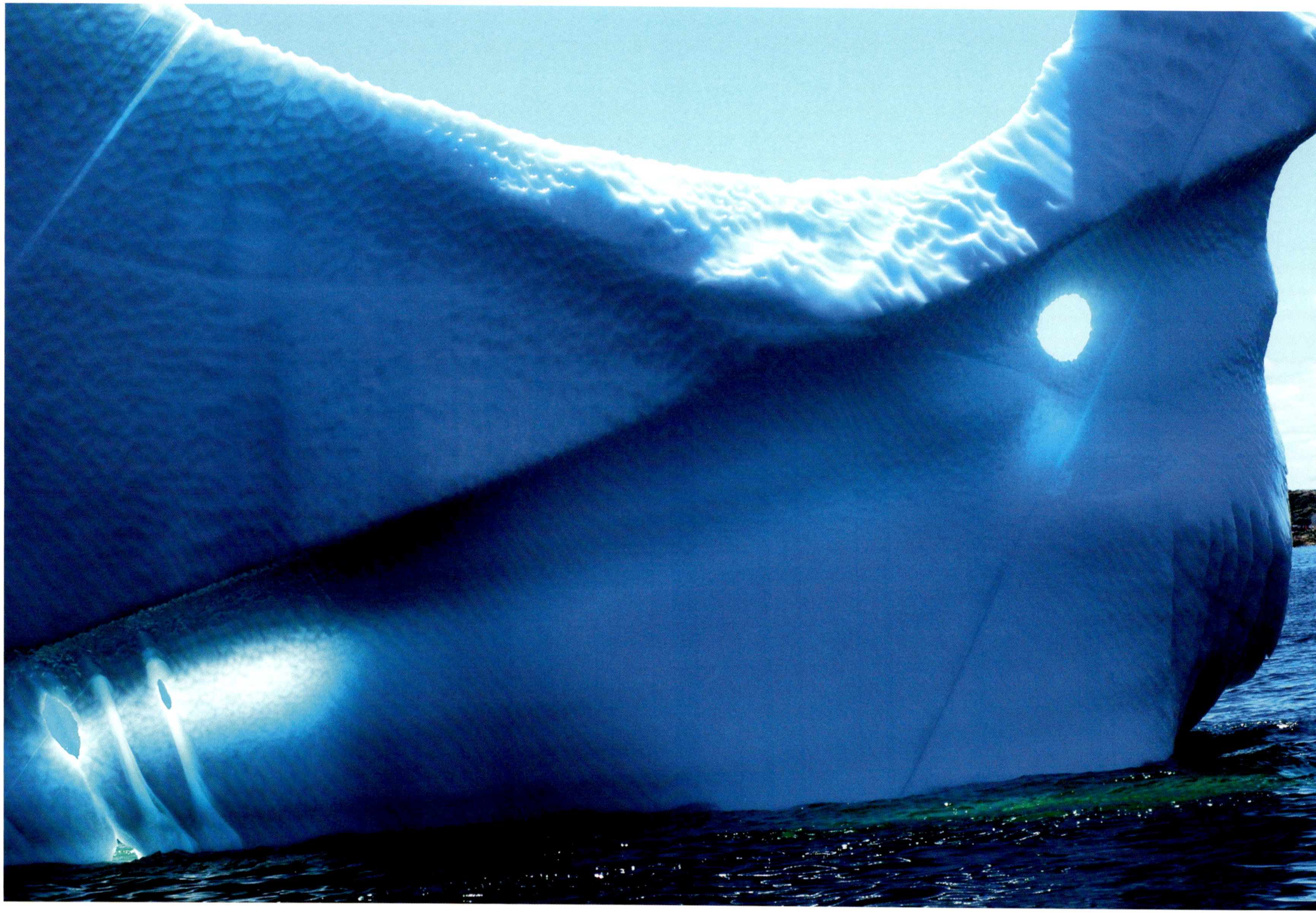

Viking Medley

Kevin

Early Viking explorers, Thorvald and his son Eric the Red fled Norway and Iceland respectively as exiled murderers. It is interesting to note that their explorations while in exile followed a route that included: southwest Greenland, Baffin Island, Labrador, and "Vinland"... similar to the route travelled by the icebergs we sighted in the communities surrounding L'Anse aux Meadows today.

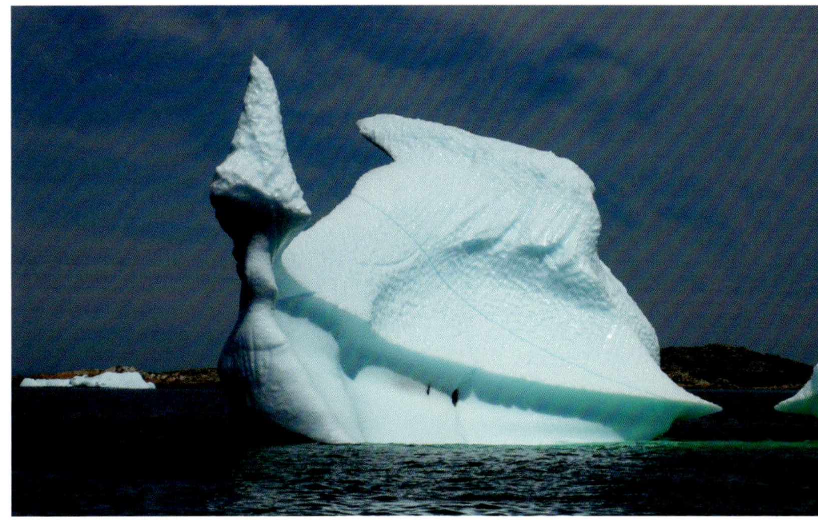

In fact, we couldn't help but think that the icebergs here today were likely looking down on the Viking ships as they passed Greenland some thousand years ago, and with that time representing approximately 1/10 of the iceberg's life cycle and approximately 500 human generations . . . a sobering and thought-provoking realization of our scale in terrestrial timelines.

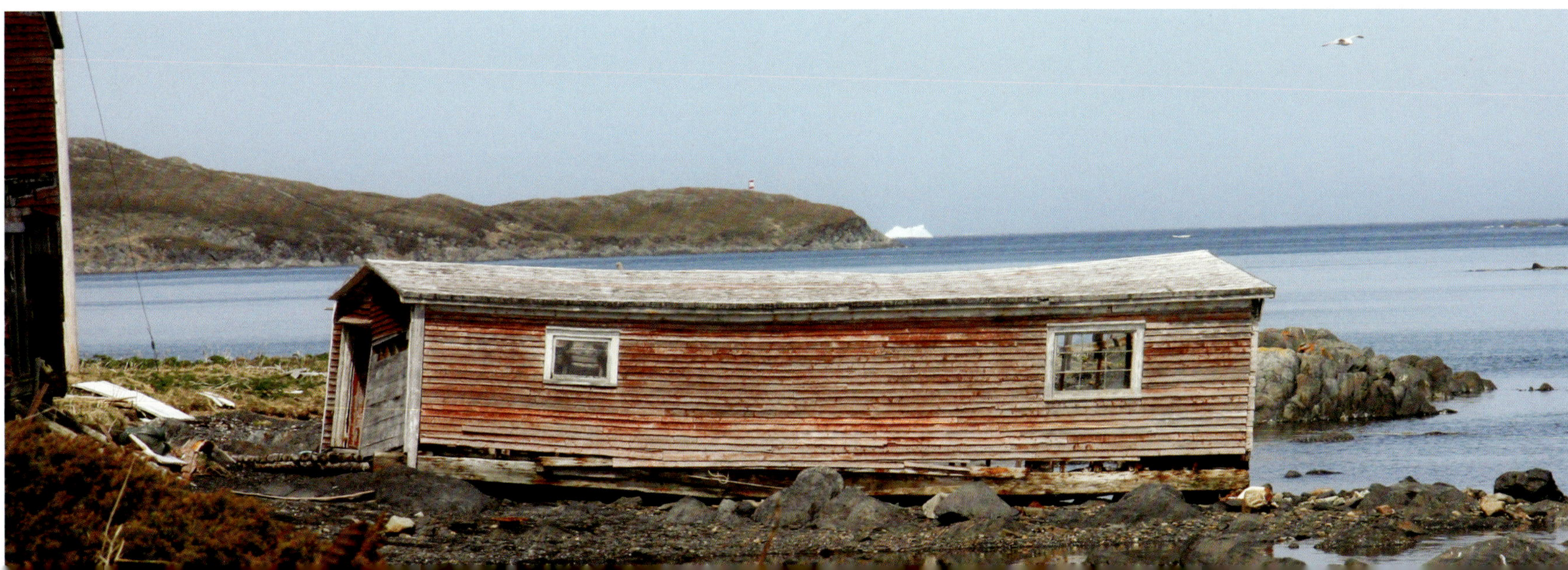

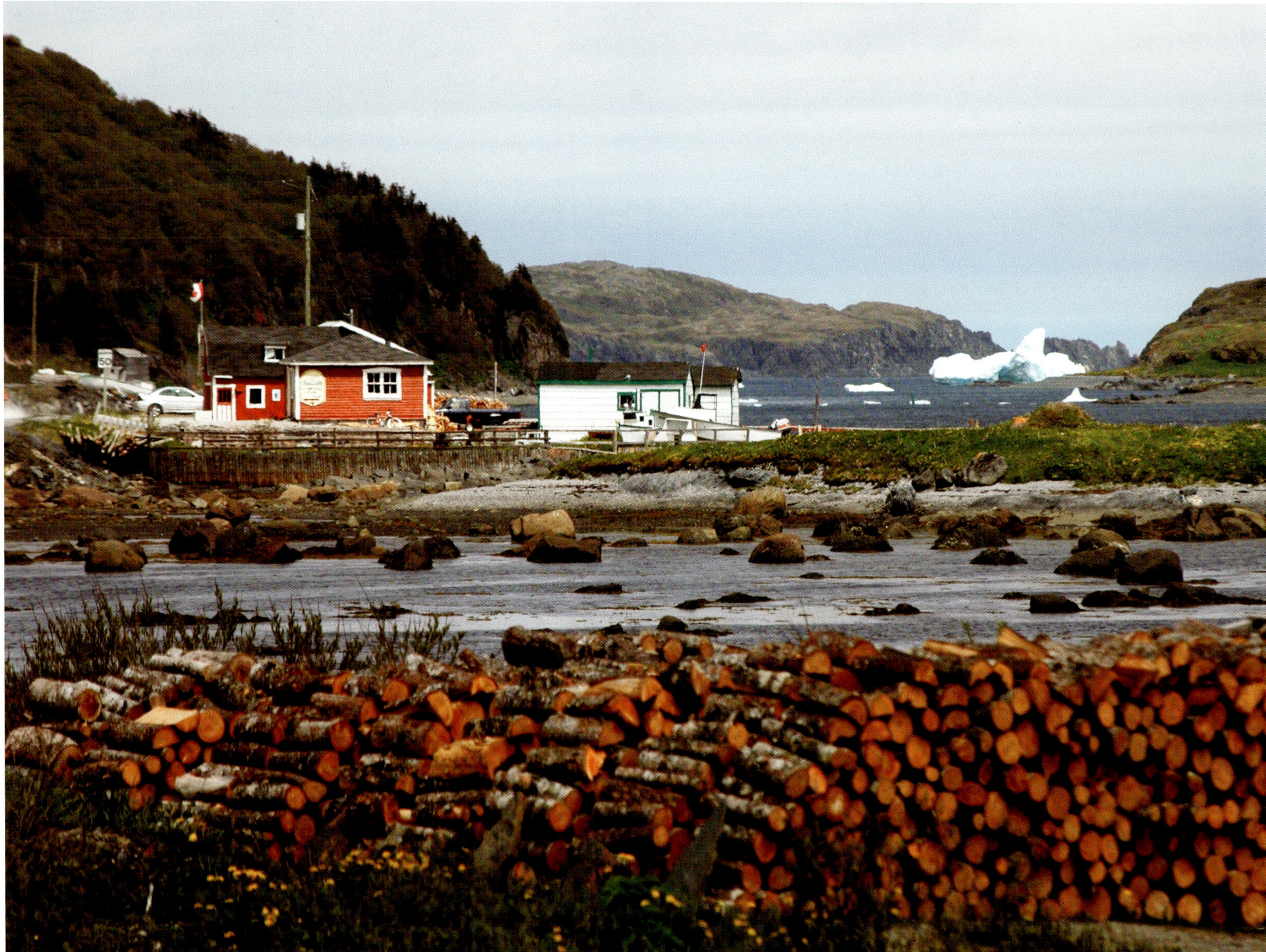

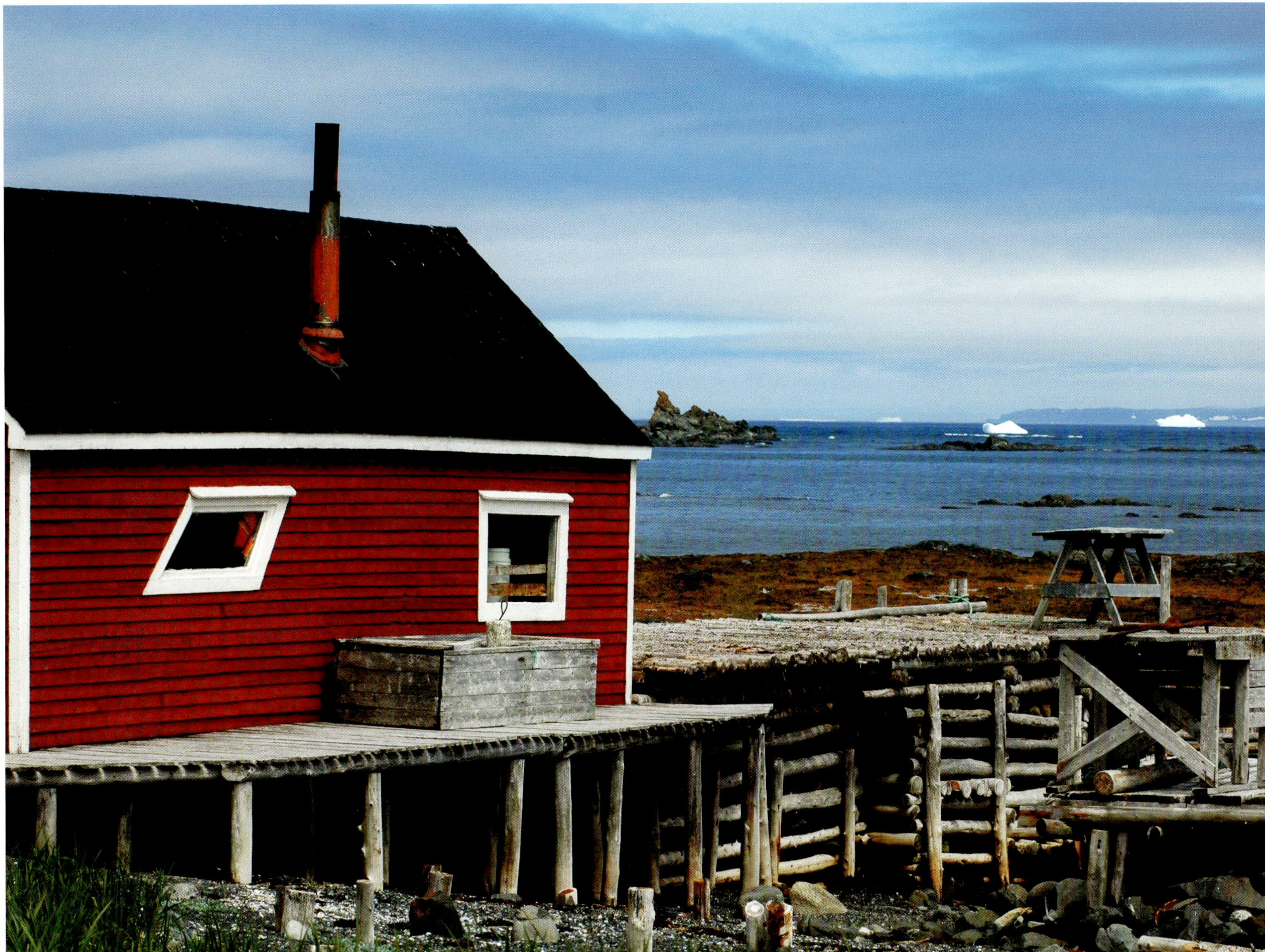

The Saga Continues

Steve

St. Anthony may have the longest iceberg season on the island of Newfoundland. Bergs begin showing up in late winter as the Labrador pack ice moves down. The spring sees the waters in the area almost continuously covered with varying amounts of pack ice with icebergs tangled up in it. As the spring wears on, the sea ice begins to fade but bergs continue arriving, some remaining, others moving on to more southern locales. Frequently iceberg fragments last well into July, but with sea temperatures rising they deteriorate rapidly.

I visited the L'Anse aux Meadows National Historic Site and Viking interpretation centre and spent the better part of an afternoon reading chunks of the Norse sagas. Zeroing in on the parts pertaining to expeditions of discovery and settlement, it was interesting to note that sailing conditions were typically described; the new land, its bounties and peoples were also detailed. However, there was no mention of icebergs or description of the treacherous ice conditions that must have prevailed for much of the time. It made me think that these conditions must have been so very typical for these far-northern dwellers that they were unnoteworthy or dropped by the saga editors . . .

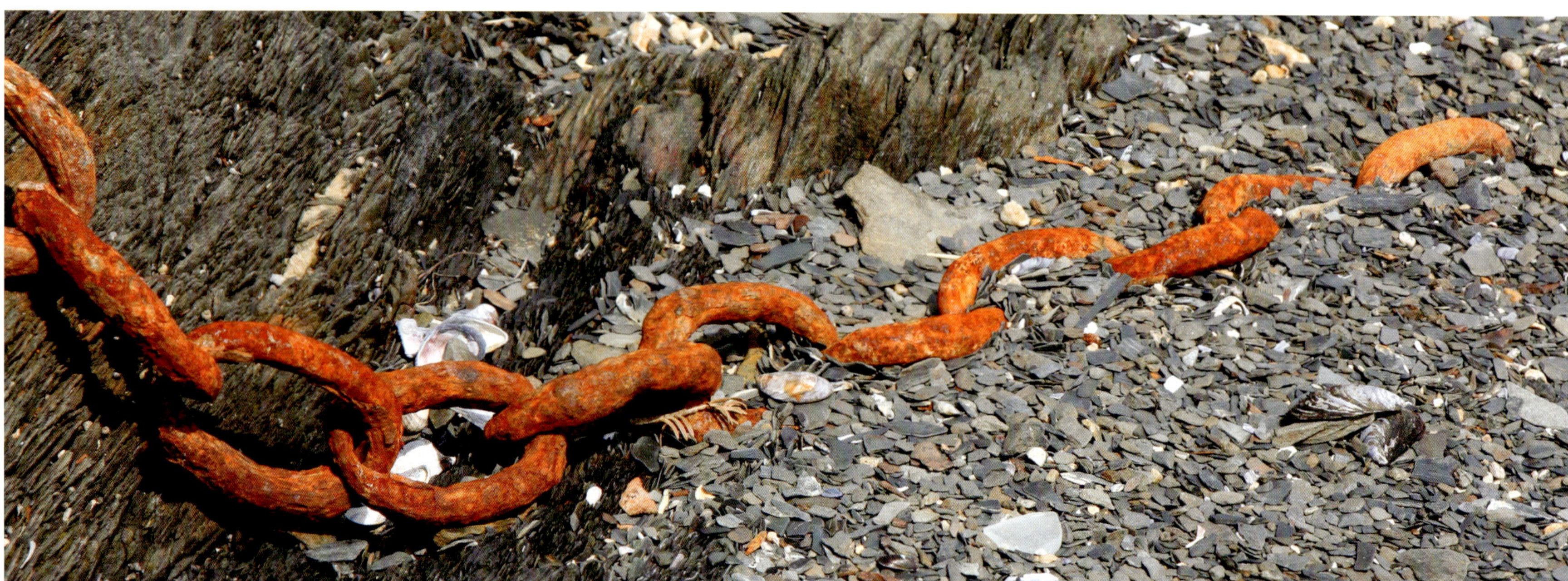

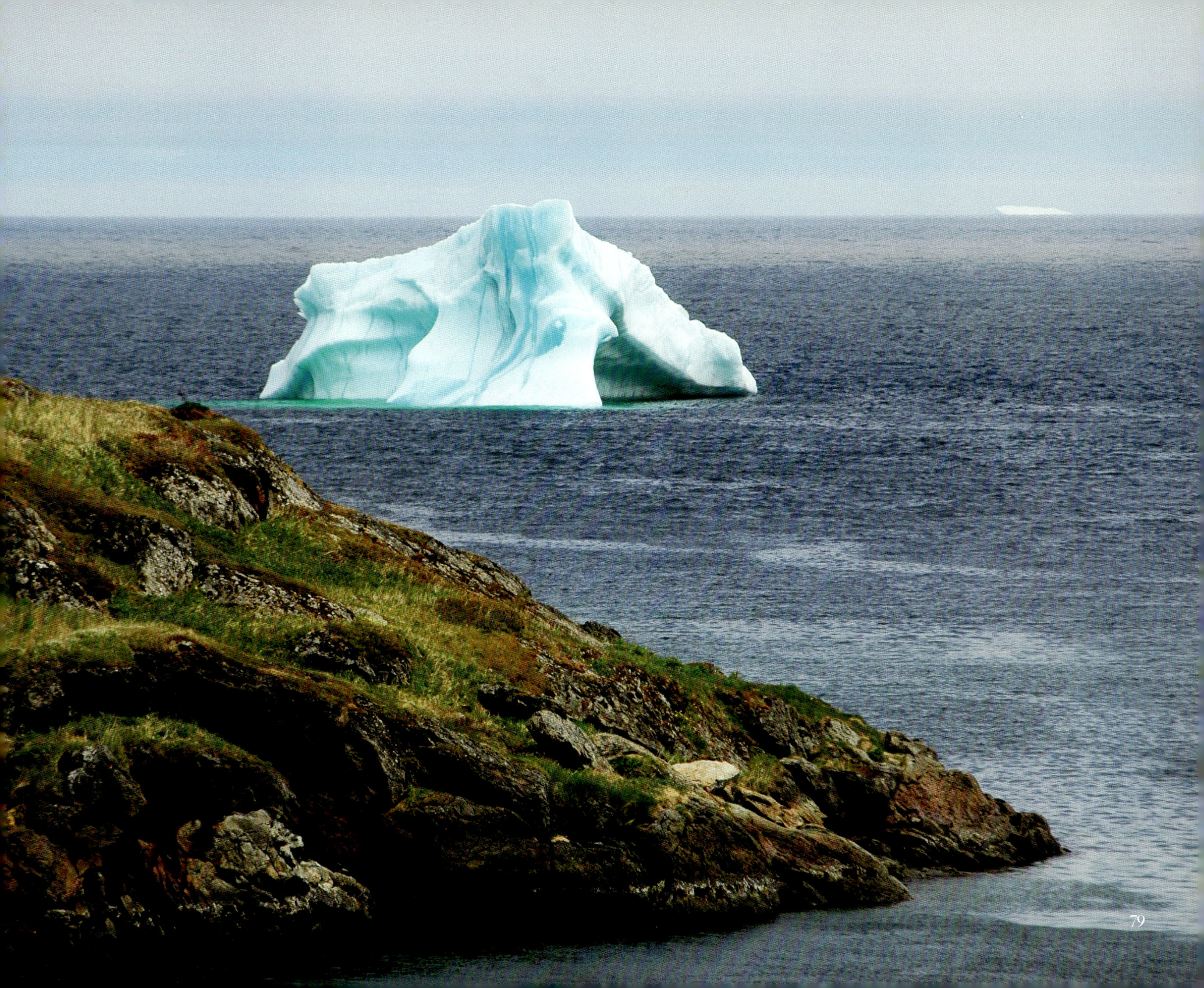

Floating rocks . . .

Steve

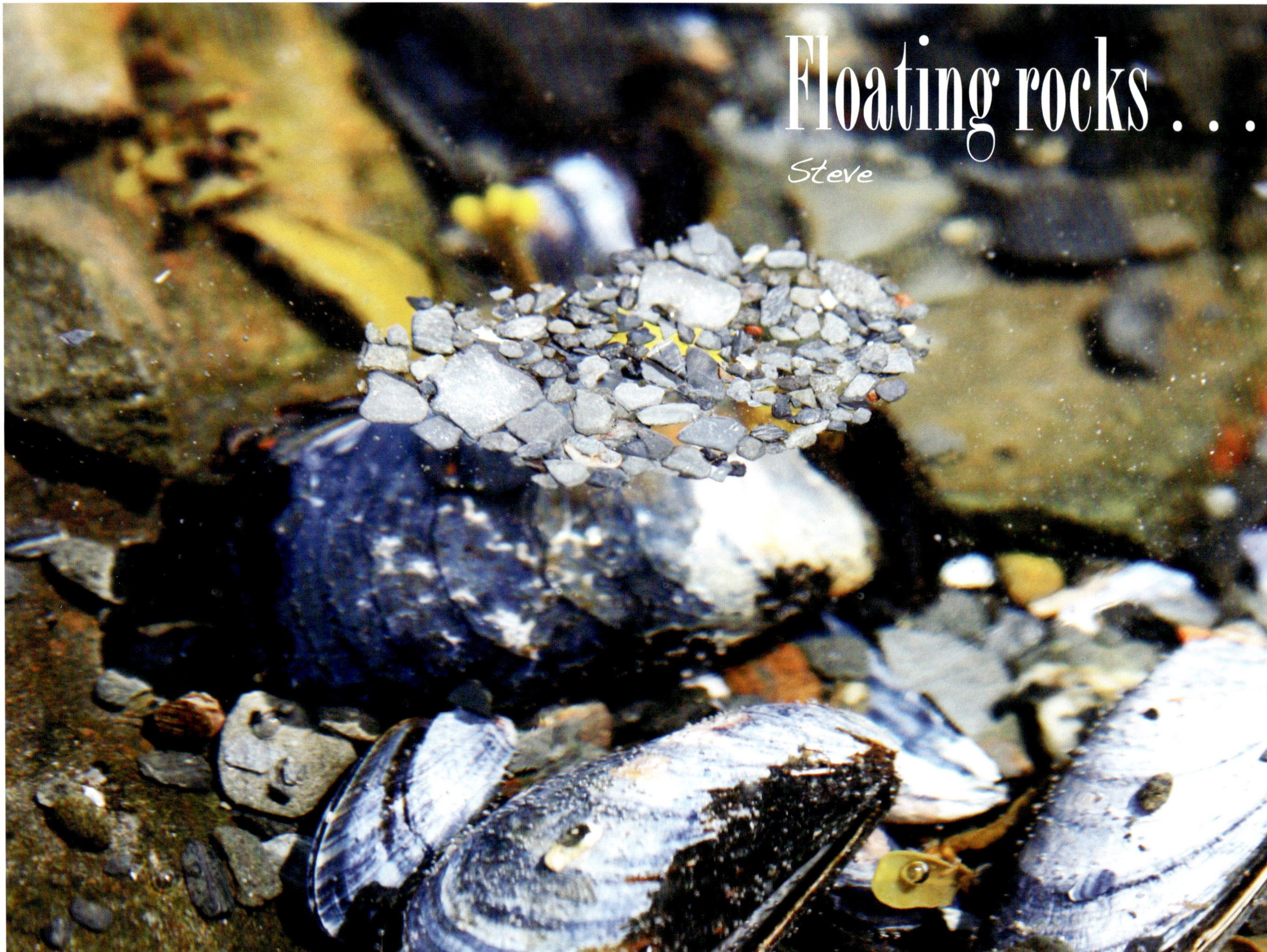

. . amongst other things.

QUIRPON

Ice and wood float because they are less dense than water; a steel ship floats because it is hollow and water is prevented from flowing inside to fill the void. So here I am looking at a small cluster of floating rocks, wondering "Now, how do I explain this?"

Seems that the local beach is made of shale fragments that in some cases are so small and thin that the surface tension of the water around the perimeter is sufficient to keep them afloat. For this to occur the rock pieces must be bone dry, at least on the upward-facing side, and the water must be so calm that an incoming tide can gently lift the stones without agitation.

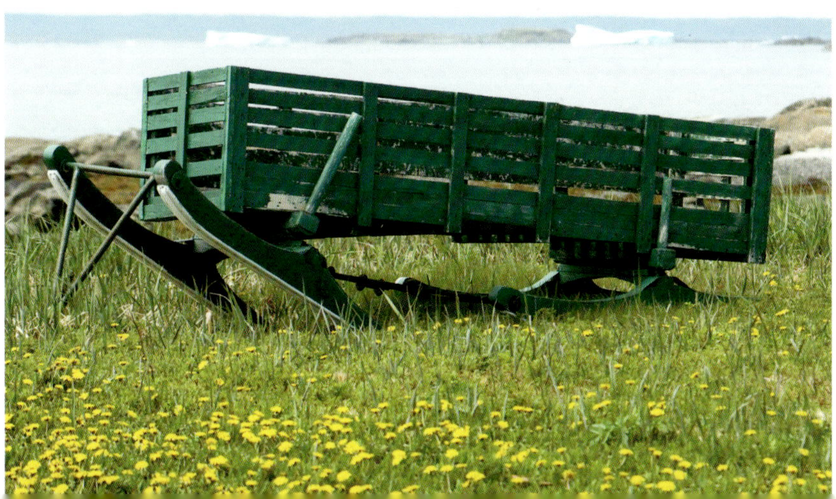

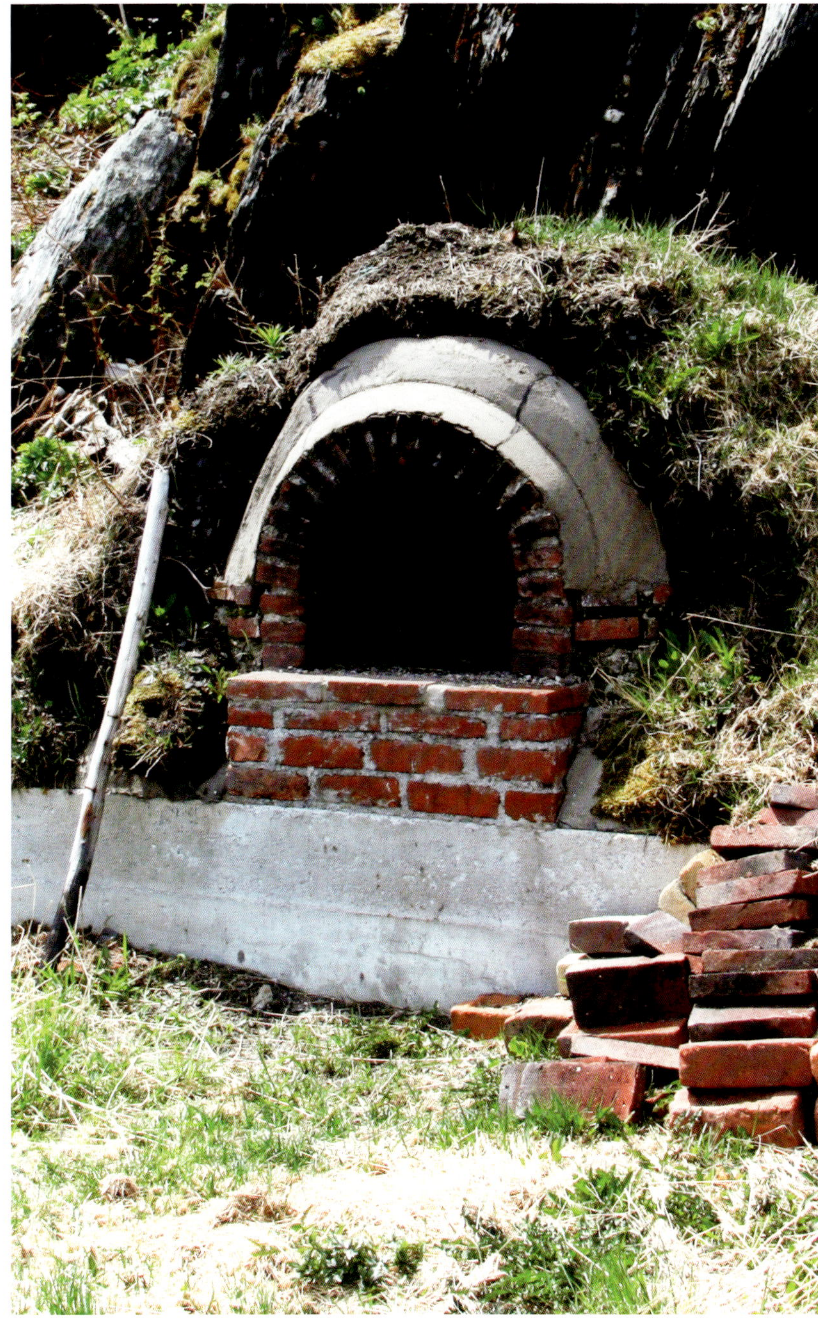

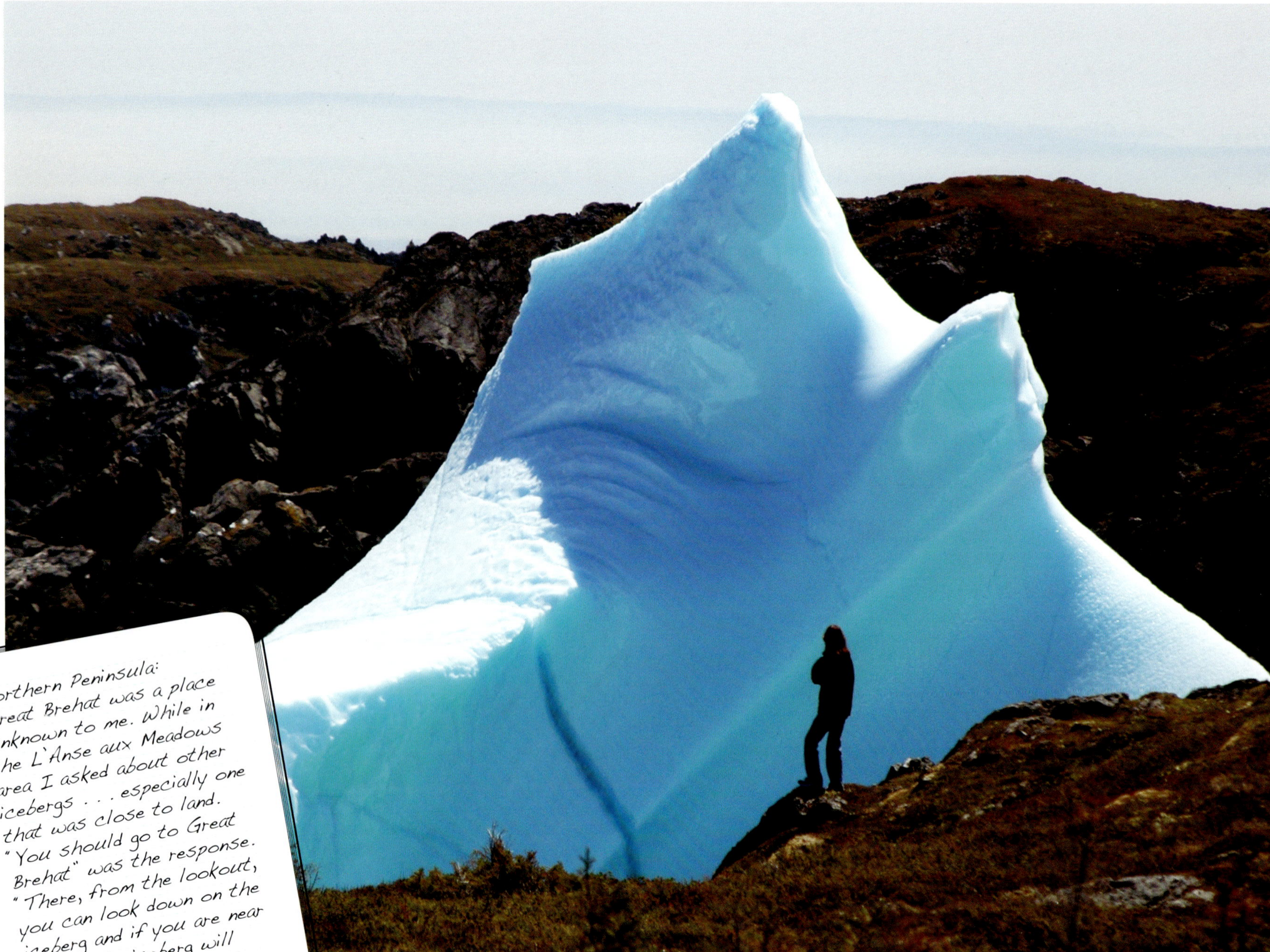

Northern Peninsula:
Great Brehat was a place unknown to me. While in the L'Anse aux Meadows area I asked about other icebergs . . . especially one that was close to land. "You should go to Great Brehat" was the response. "There, from the lookout, you can look down on the iceberg and if you are near shore, the iceberg will look down on you!"

Great Brehat Beauty Queens

Kevin

The easy ten-minute walk to Flat Point Lookout in Great Brehat took us about forty-five minutes. Halfway to the lookout, we were scanning every direction until, to the southeast, a bony mass of white ice seized our gaze. From our position, we could see no water. Out of place, it appeared to be part of the land. From my vantage point this shot called for a long lens. Dancing a short distance over the barrens to the northerly elevated side of the trail, I found the spot for the shot. By the time I was ready to shoot, a beautiful woman had appeared atop the precipitous cliff near the iceberg, her dark clothing in full contrast to the whiteness of the berg. Her shapely stance challenged the iceberg's finery like two enamoured beauty queens looking at one another and being the best of friends.

From her vantage point this lady could look down upon the bowl of ice supporting the spiny berg's back hovering overhead, the iceberg faintly resembling a giant conch shell. She could also see a second iceberg stranded in the tickle, all invisible to me. Here in the narrow Great Brehat tickle were two grounded icebergs, which were likely once joined, forming an archway and grounded on the shoal east of Flat Point in early June.

Thinking of the changes, it seems that such huge transformations appear effortless and natural, like water to ice only in reverse.

P.S. *Find out the meaning of the French word "Brehat."*

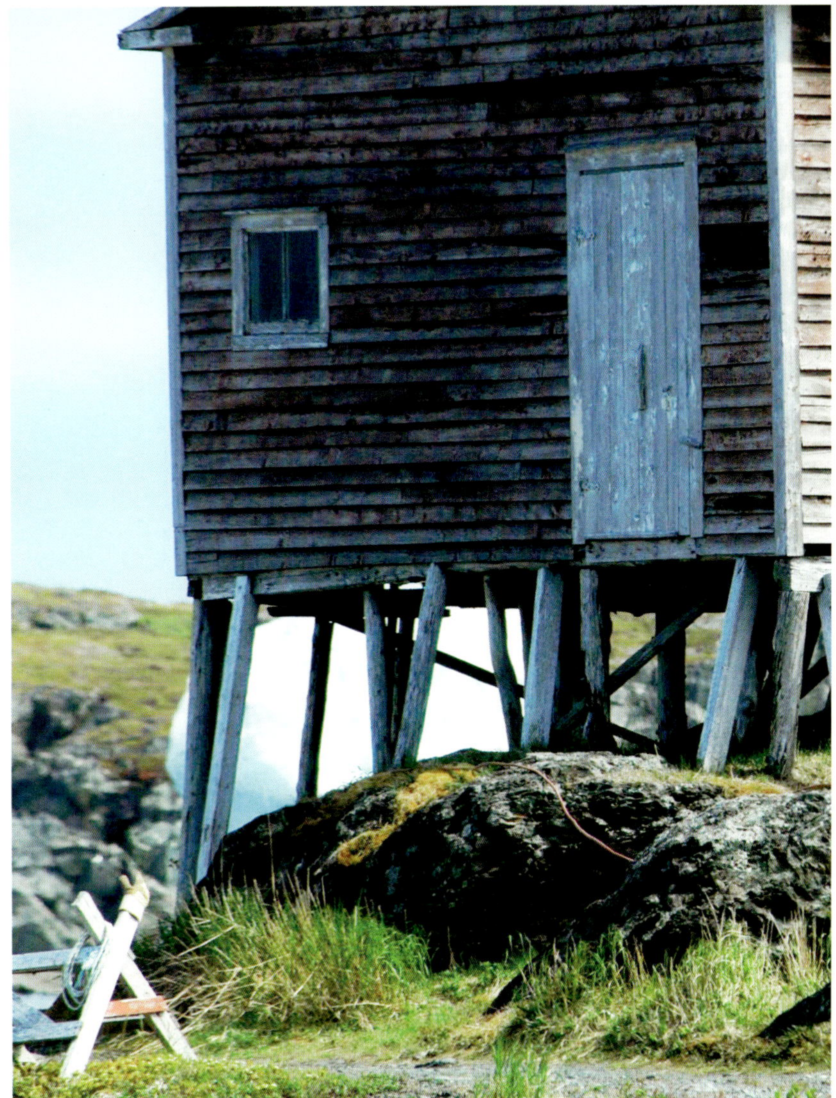

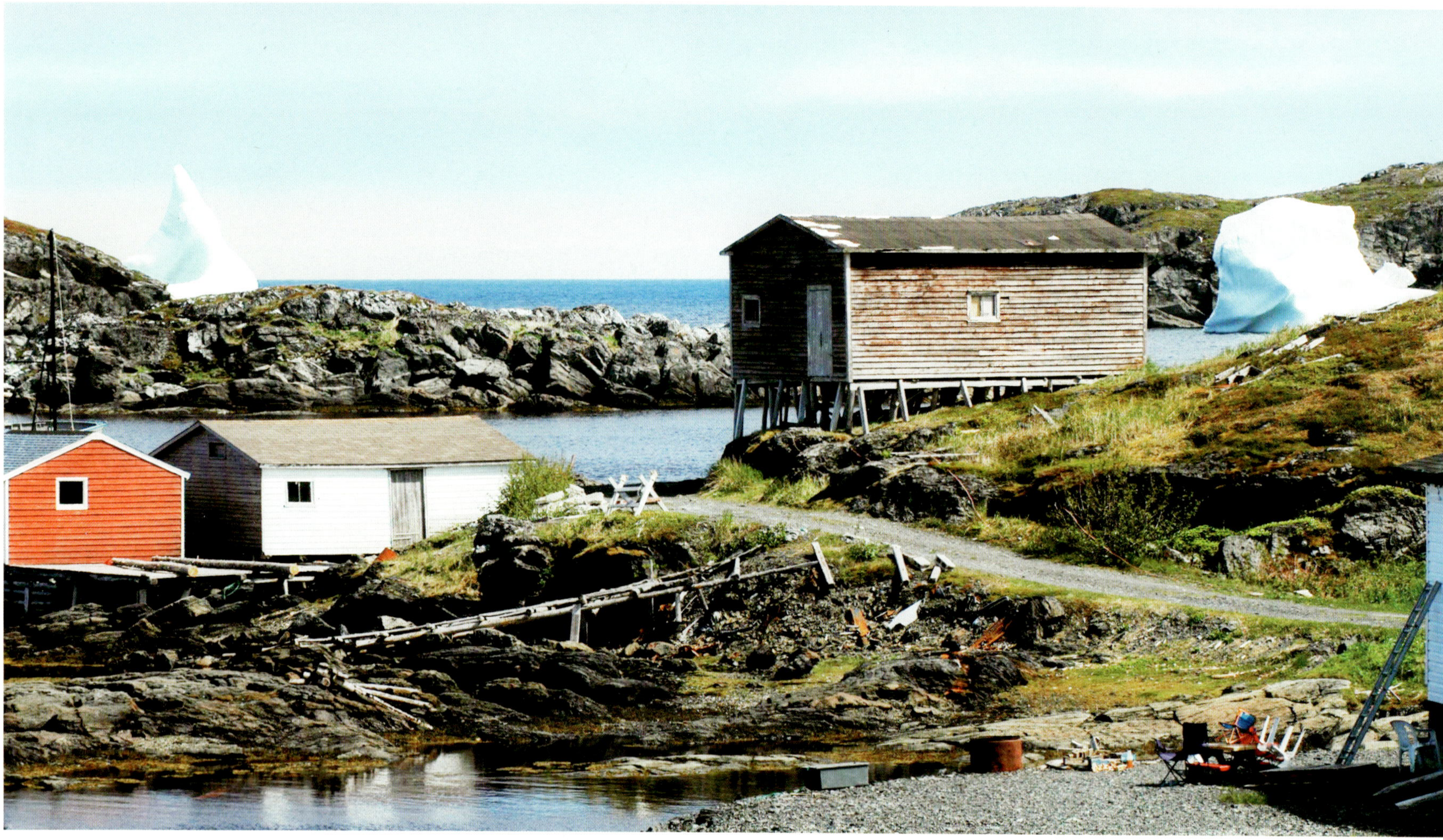

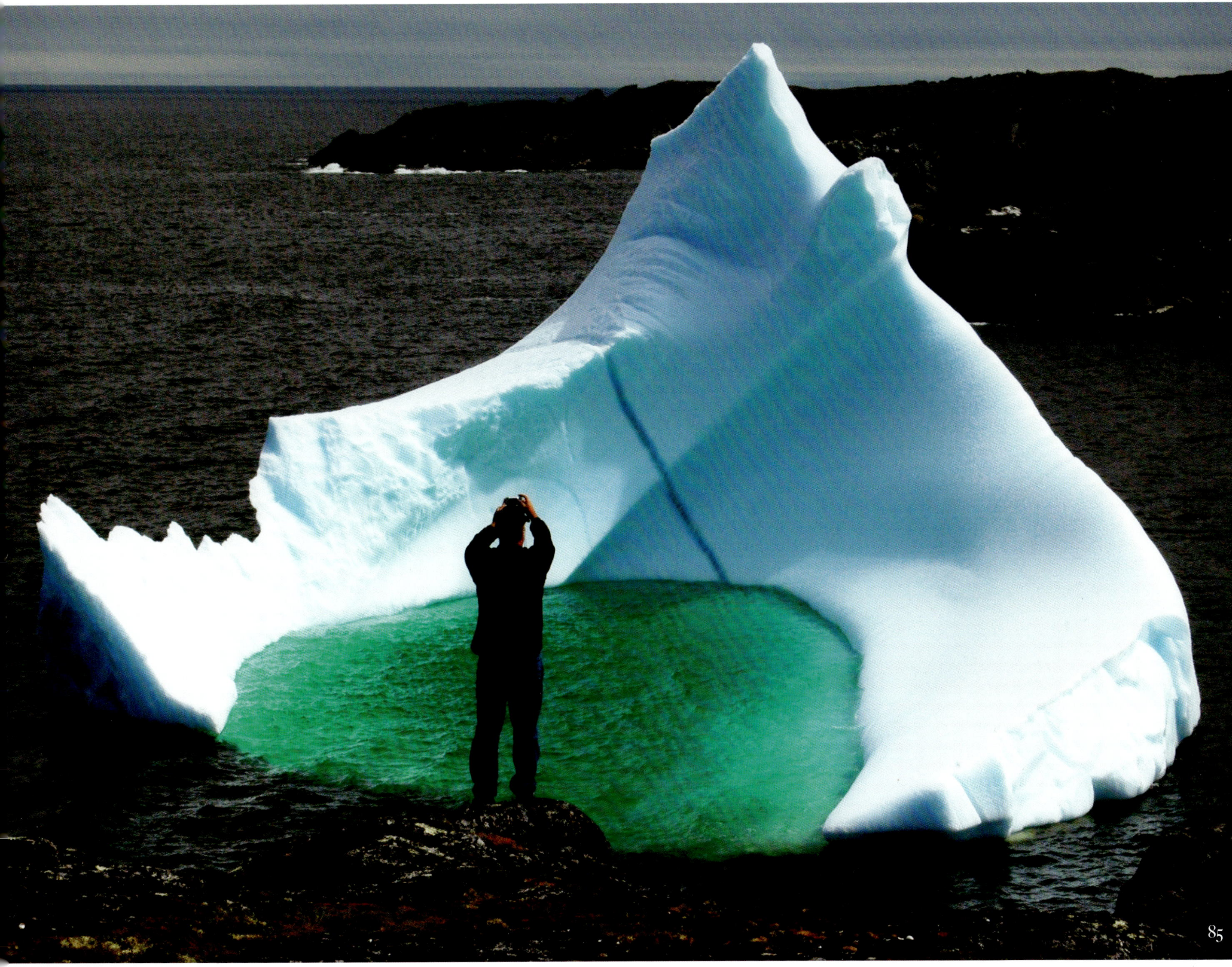

Anthony: Patron Saint of Lost Things

Steve

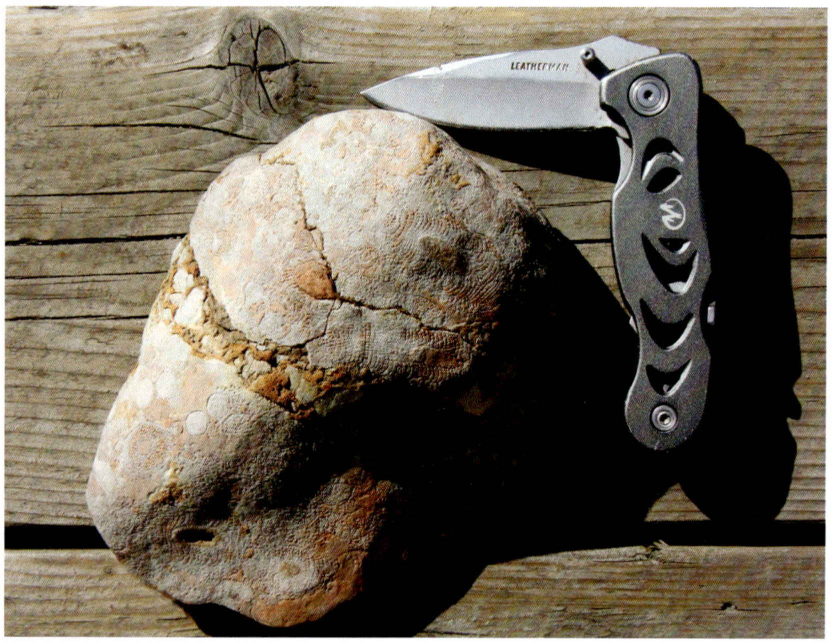

The best fossil I now have in my modest collection is an exquisite view of ancient animal life on the seabed. The various aspects from which the sponge-like organisms can be seen and felt are really quite remarkable. Stranger than fiction is the fact that I found this specimen amongst a pile of rock debris on the floor of an untidy rental car on a recent iceberg trip to St. Anthony . . .

I have heard about strange things being seen, or found, IN icebergs. Stories have been told of old aircraft, shipwrecks, and even skeletons being sighted inside the ice of a drifting berg. Of course I don't have any evidence that any of these stories are true, but reflecting on my own experience, I'd best keep an open mind.

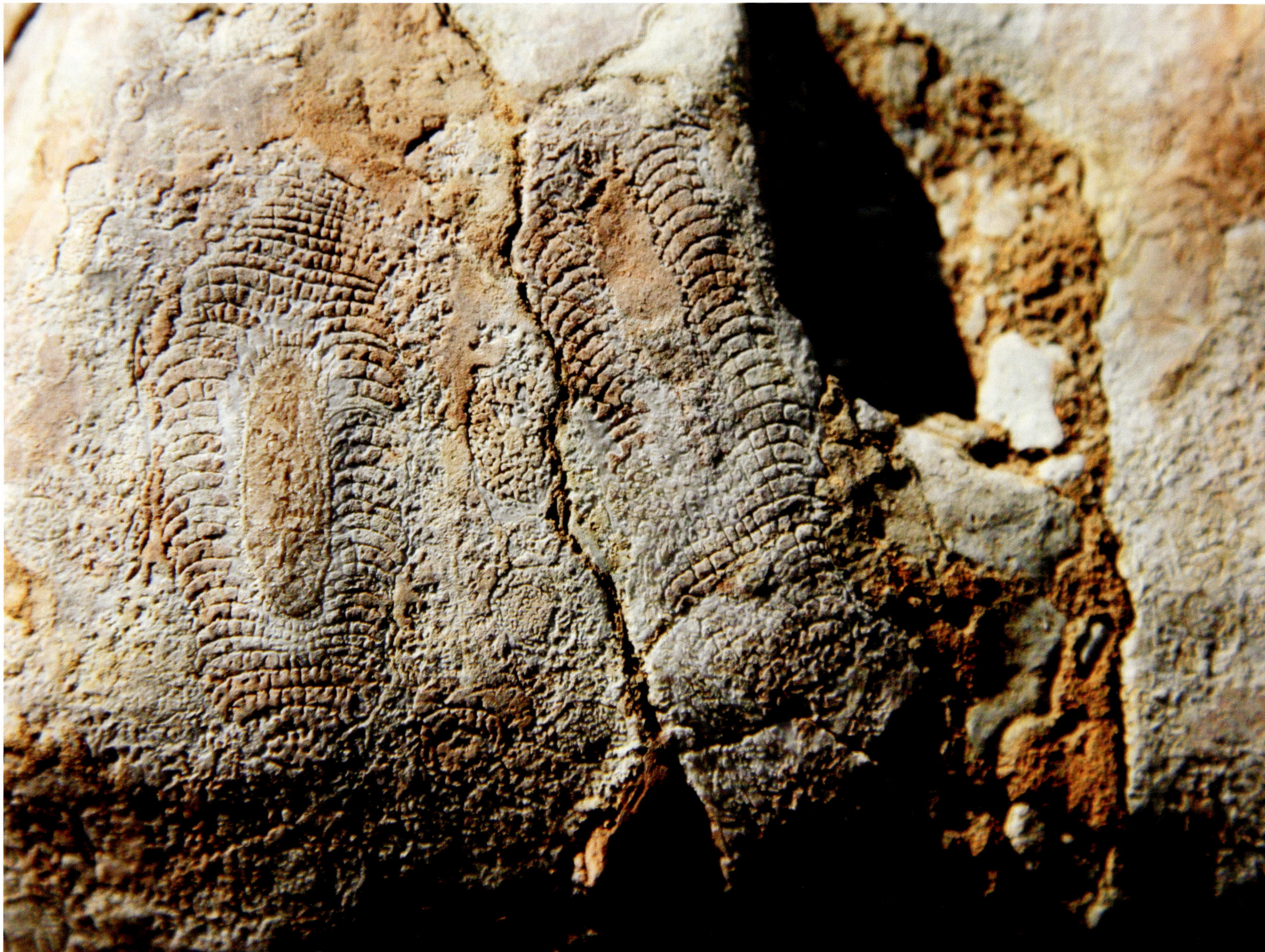

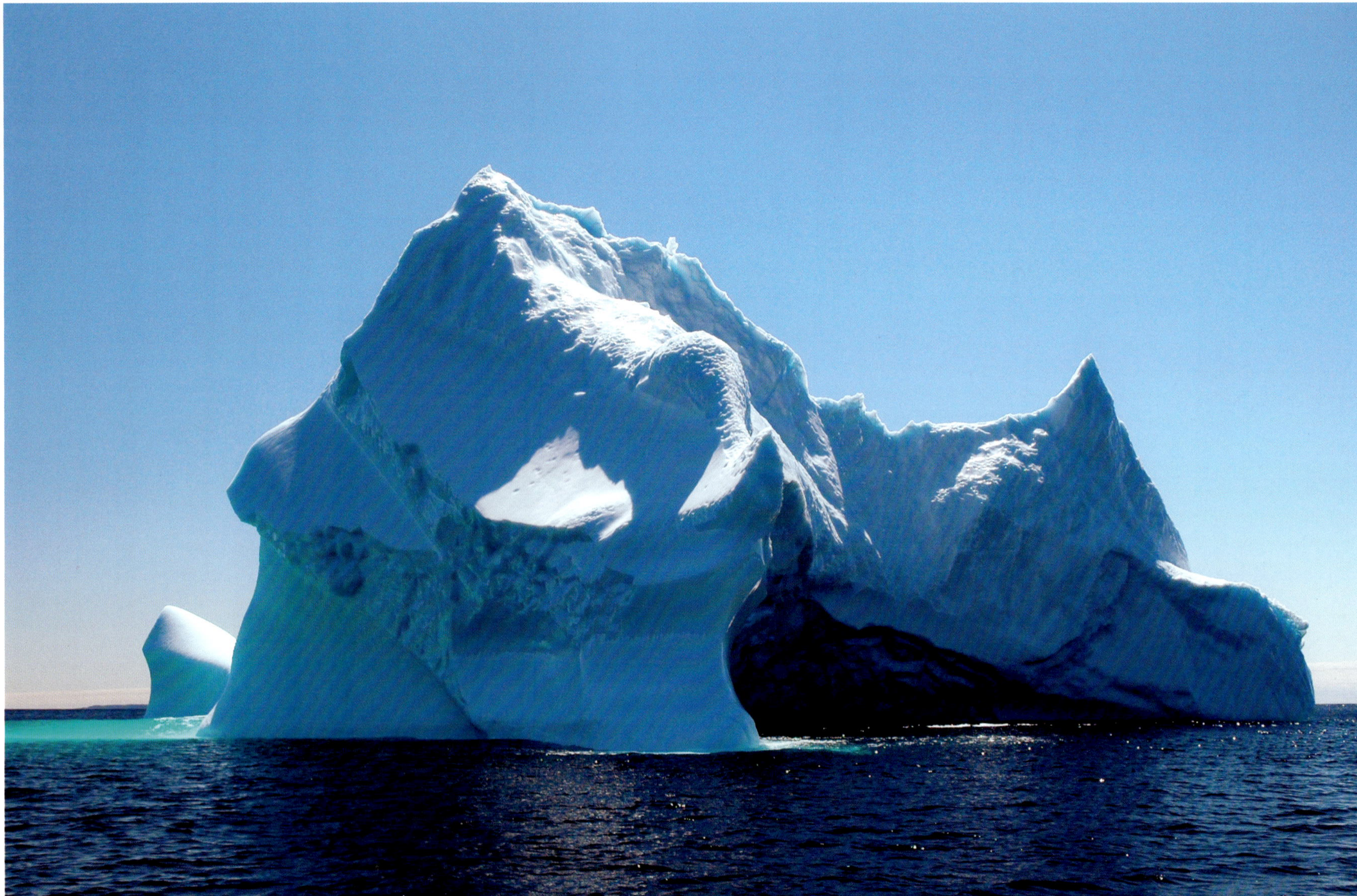

"Ass Over Kettle"

Kevin

Heightening my gaffe was, once on the road, the radio forecast for White Bay earmarked rain and showers . . . less than ideal conditions to photograph the only iceberg on Newfoundland's northeast coast. Bergs were scarce this year and my desperation for iceberg shots seemed to be driving me over the edge of rational behaviour!

Despite the forecast, blue skies welcomed my arrival to Conche at 10:30 a.m.

A moderate southwesterly wind gently rocked the boat as Chris Kearney and I made our way to the iceberg grounded in Maurice's Cove. My first sighting of the berg left me sadly disappointed, prompting a knot in my gut. It looked so small! Hence I photographed each small bergy bit en route to the mother berg.

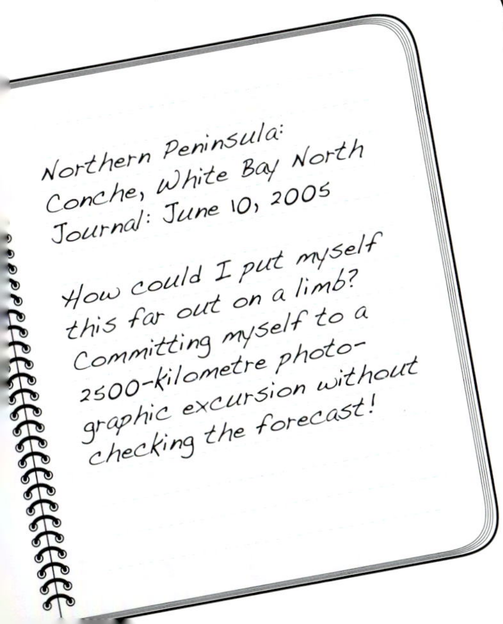

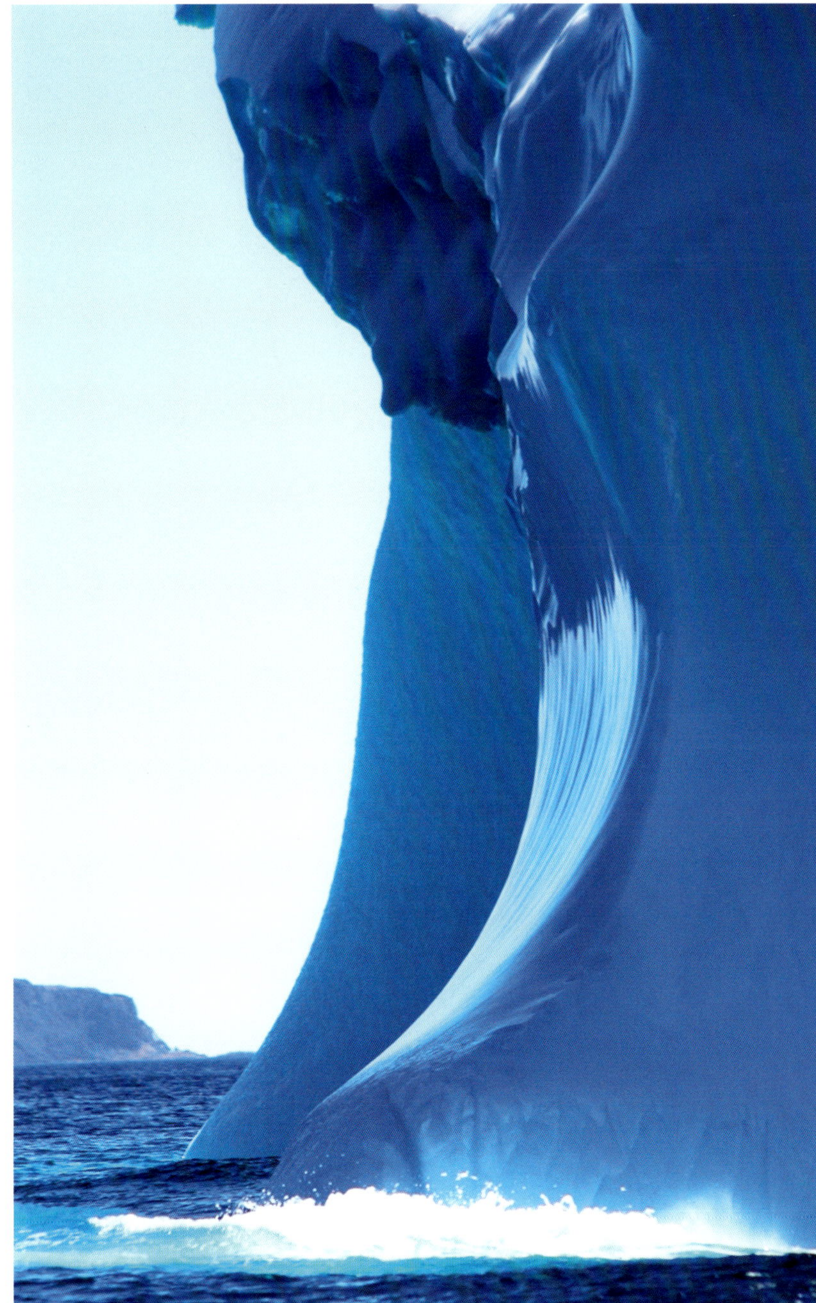

The closer we came to the iceberg, the bigger and more interesting it appeared. In fact, it became the smorgasbord king for that season because of its diverse features that included: a face, crown, pillars, and even phallic-like ice sculptors. At one point Chris (the boat's captain) sent me "ass over kettle," flying over the bow seat, landing on my back into the boat's middle bay as he sped away from the thunderous roaring noise emanating from the iceberg's core . . . the gunshot-like sound being a normal precursor to a berg's fracturing.

As I complete this journal entry I am filled with delightful feelings from my visit to Conche. Not only did the iceberg exceed my expectations, but the community's hospitality and historical charm like that of fifty years past! The nineteen-kilometre dirt road to the community included thirty-four blind hills and turns . . . a step back in time!

P.S. *Note to self: Mark Conche on my map of places deserving a return sojourn!*

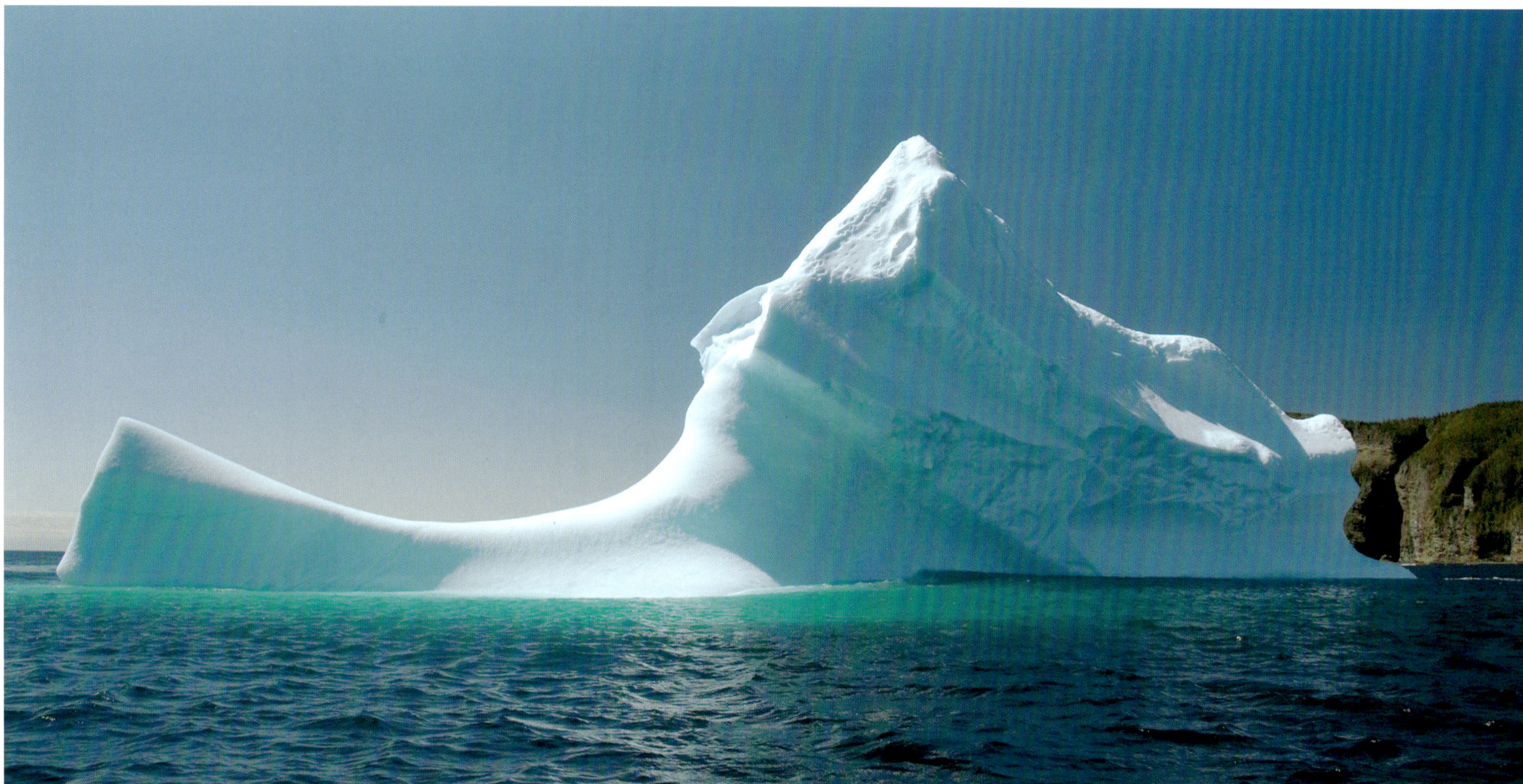

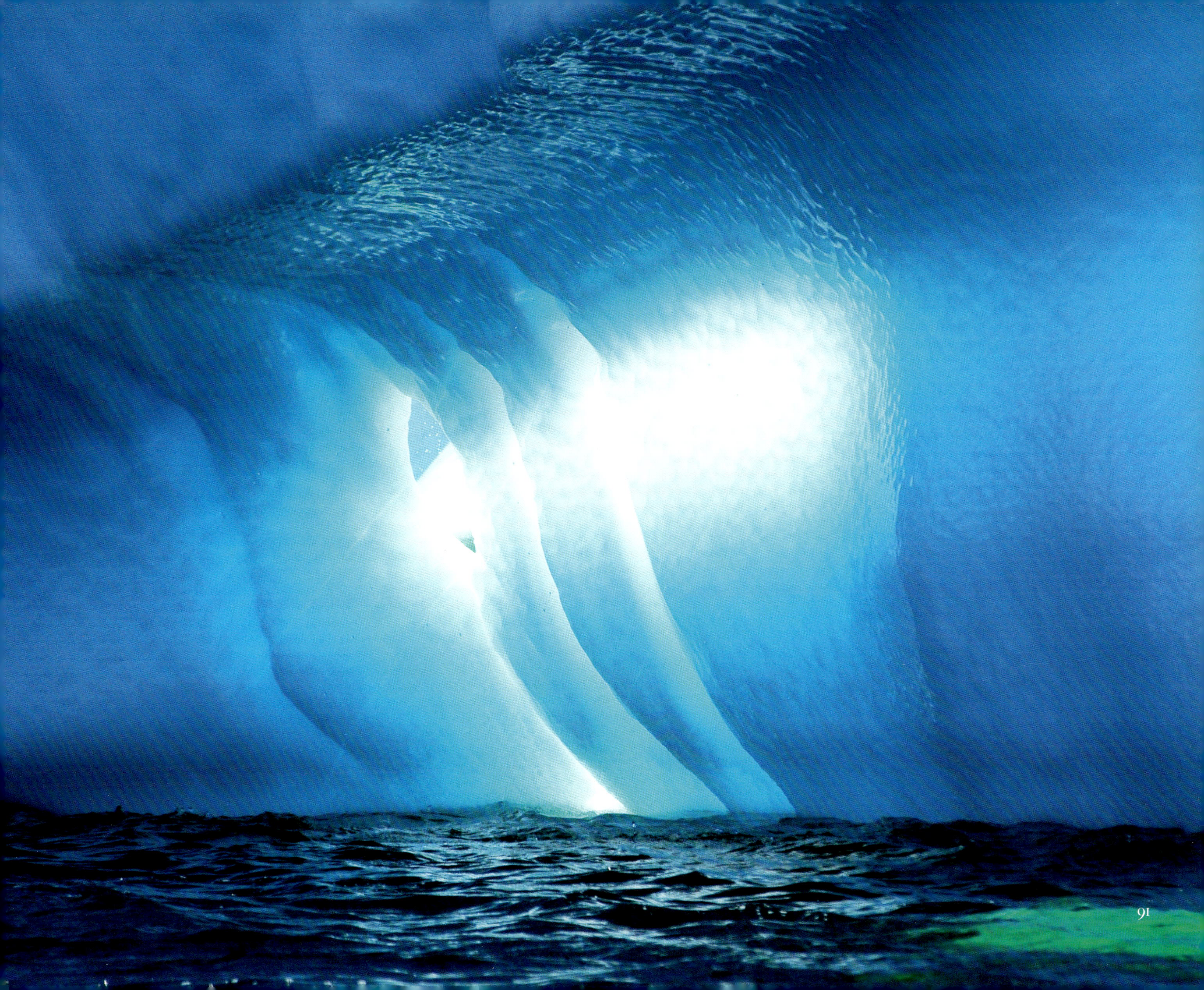

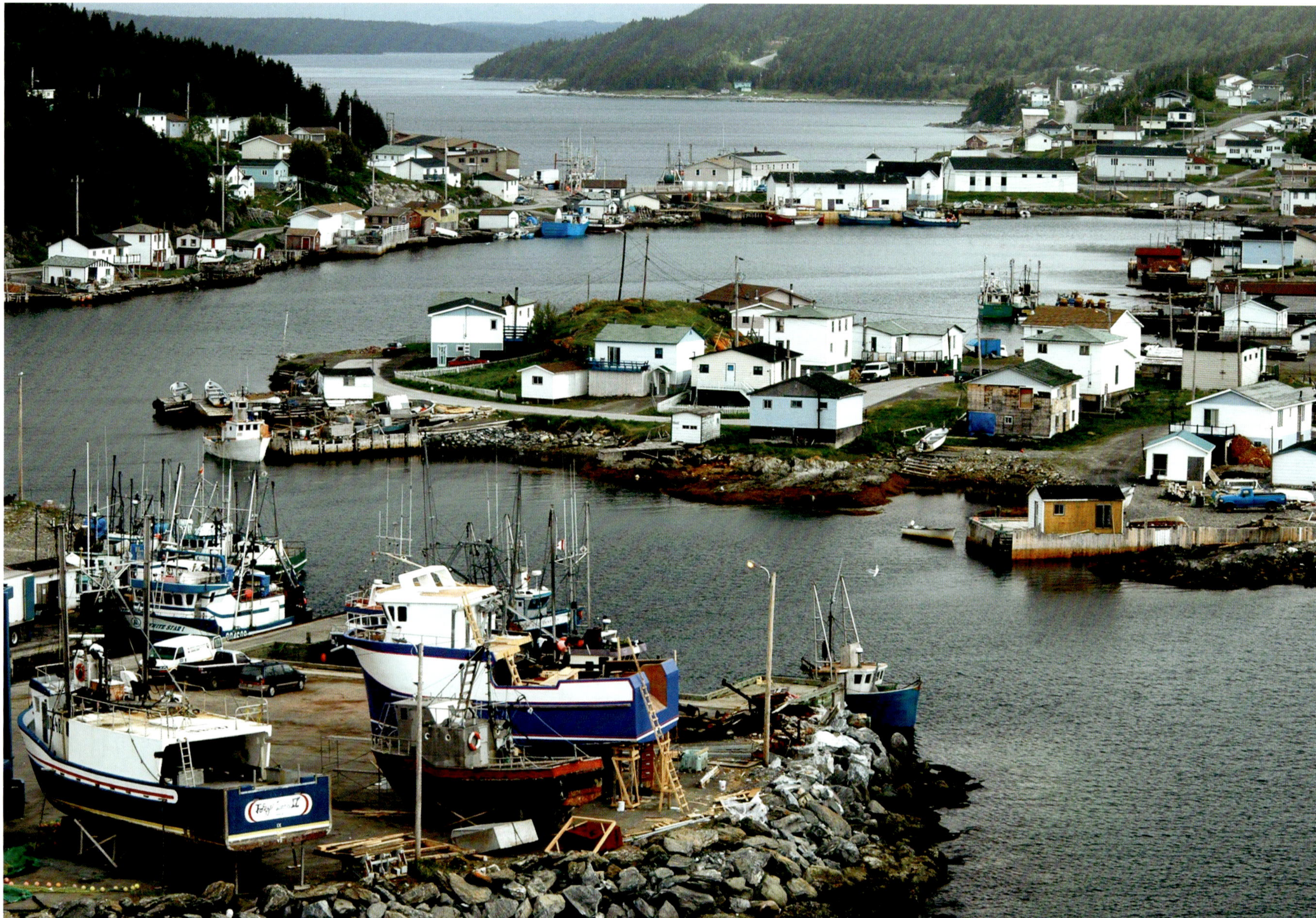

Water Haul

Kevin

With great anticipation, I arrived in Englee, 10:00 a.m. this morning after a twelve-hour drive only to find no iceberg in sight. Leaving nothing to chance, I walked the hiking trail and climbed the hilltop featuring panoramic vistas of community and ocean; still no iceberg in sight. Finally, stopping in the marine service centre store, local fisherfolk confirmed the iceberg was gone from Englee and likely moved on down the remote shore toward Great Harbour Deep. Then, as if on cue, a stranger entered the store, and listening to the iceberg chit-chat, stated: "You wouldn't believe it, only two weeks ago some feller drove all the way from St. John's to Conche (twelve-hour drive) just to take pictures of an iceberg." Sheepishly, I replied: "That was me!"

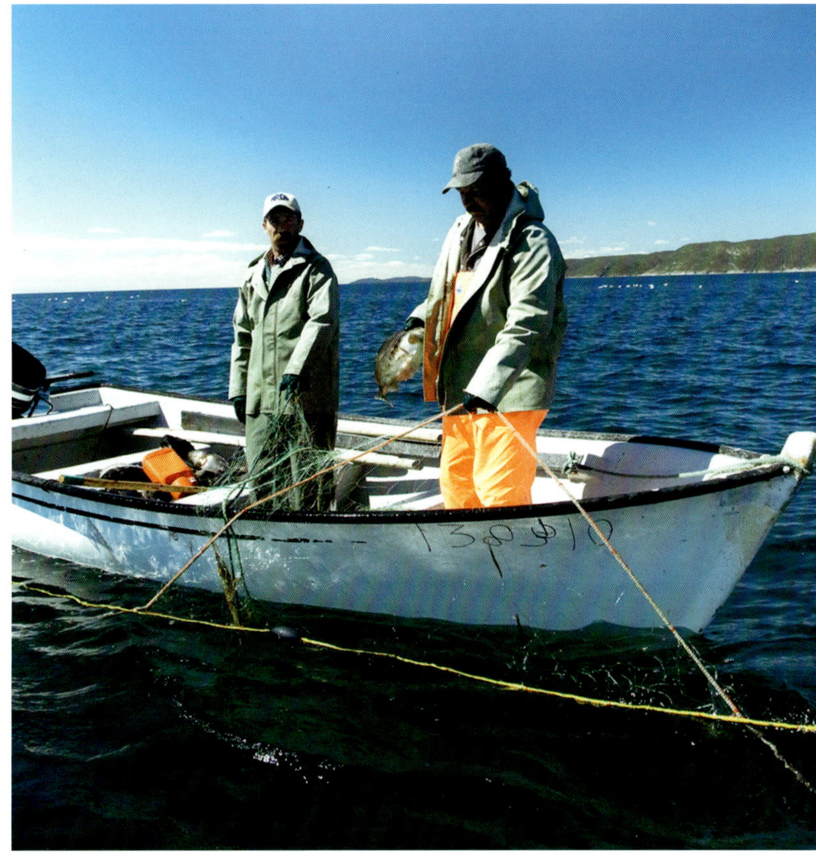

Saturday June 24

A call to Englee two days ago confirmed an iceberg in the community.

P.S. *Three years later, as we put together this book, we are reminded how fortunate we are . . . ten to twenty icebergs within a forty-five-minute drive.*

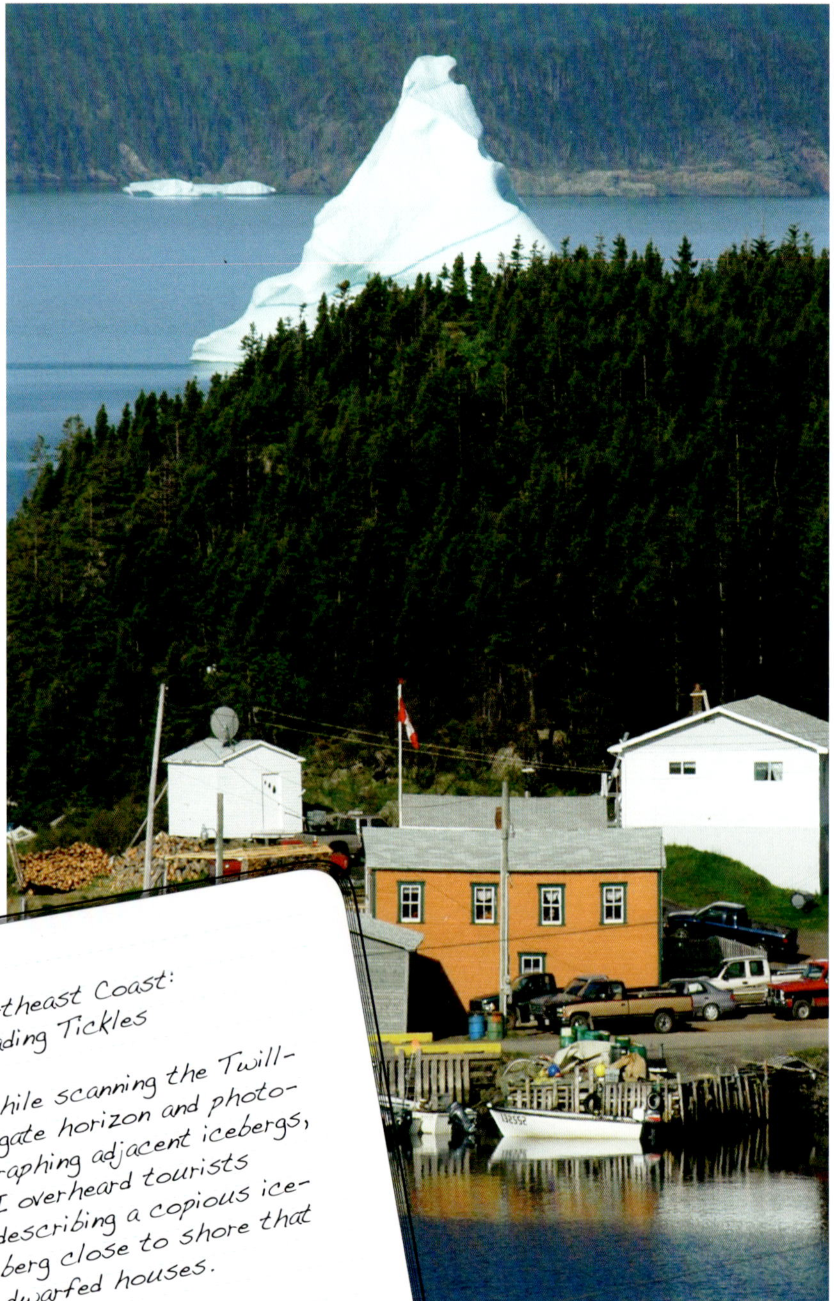

Northeast Coast:
Leading Tickles

While scanning the Twillingate horizon and photographing adjacent icebergs, I overheard tourists describing a copious iceberg close to shore that dwarfed houses.

The "Twirlee"

Kevin

I thought to myself, the community wasn't near Twillingate, for I had spent the better part of a day in the area. After a brief inquiry, the courteous couple revealed it was in Leading Tickles and described the many details of the fascinating iceberg. Although Leading Tickles was not on my itinerary, based on their description my excitement heightened; I decided to take the chance on the two-hour drive in the dark.

On arriving in Leading Tickles at 11:30 p.m., the single road in the community was deserted except for a couple carrying a dozen beer who politely directed me to the end of the road and the trail head leading to the berg. Following their directions, at no point could I see any sign of the iceberg that was supposed to be a part of this community. It was pitch-black! With no place to stay and no iceberg visible, butterflies bubbled in my belly. Choosing to rearrange the back seat of the car for sleeping rather than erect the tent, I settled in for an anxious and short sleep. It was less than five hours to daylight!

Awakened by fading darkness and stepping outside, my heart shrank when looking across the bay; it was nothing but water, no iceberg! The trail head's steep incline and thick black spruce forest blocked the view out the bay. From the peak it was impossible to see farther than the forest canopy covering the trail.

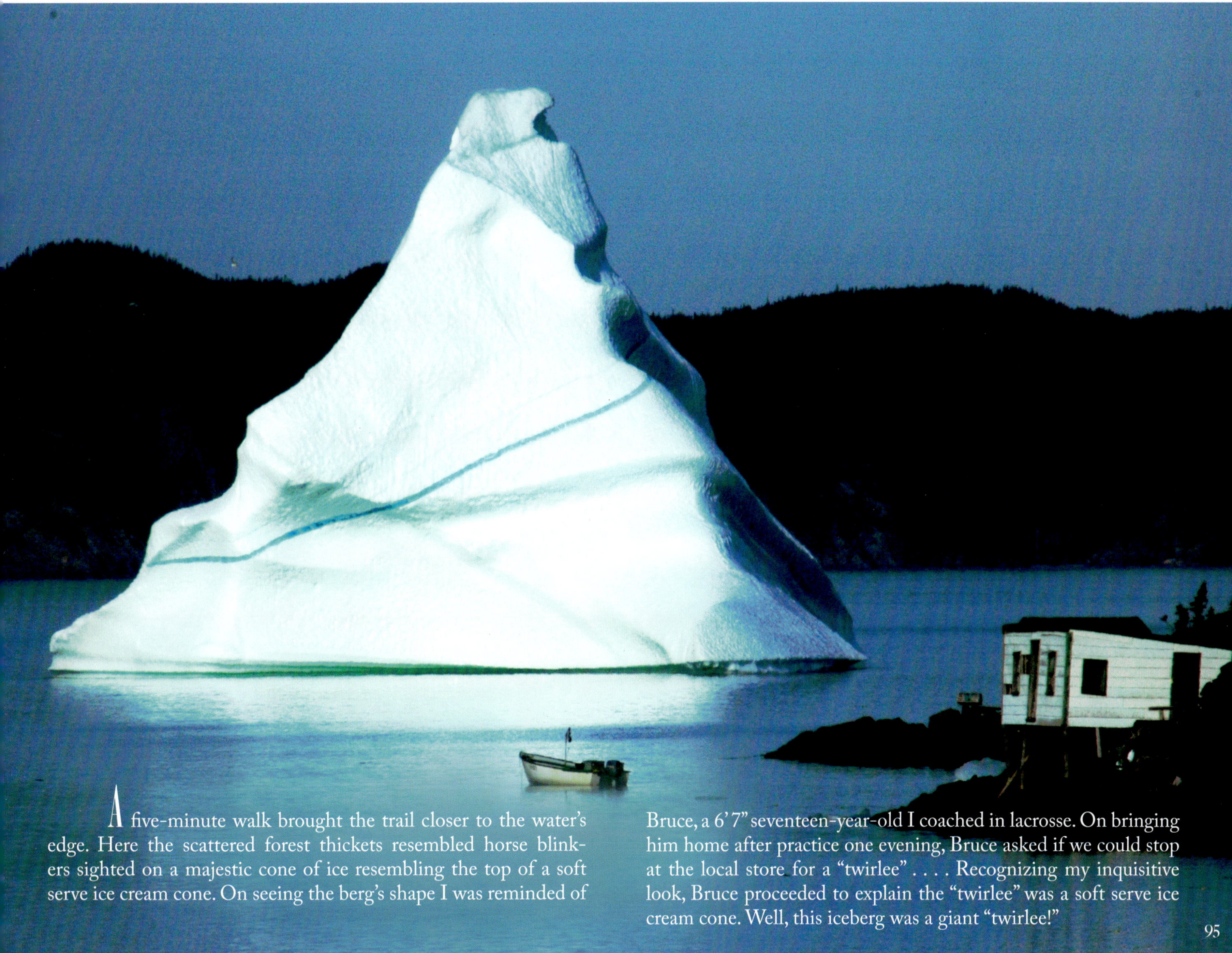

A five-minute walk brought the trail closer to the water's edge. Here the scattered forest thickets resembled horse blinkers sighted on a majestic cone of ice resembling the top of a soft serve ice cream cone. On seeing the berg's shape I was reminded of Bruce, a 6' 7" seventeen-year-old I coached in lacrosse. On bringing him home after practice one evening, Bruce asked if we could stop at the local store for a "twirlee" . . . . Recognizing my inquisitive look, Bruce proceeded to explain the "twirlee" was a soft serve ice cream cone. Well, this iceberg was a giant "twirlee!"

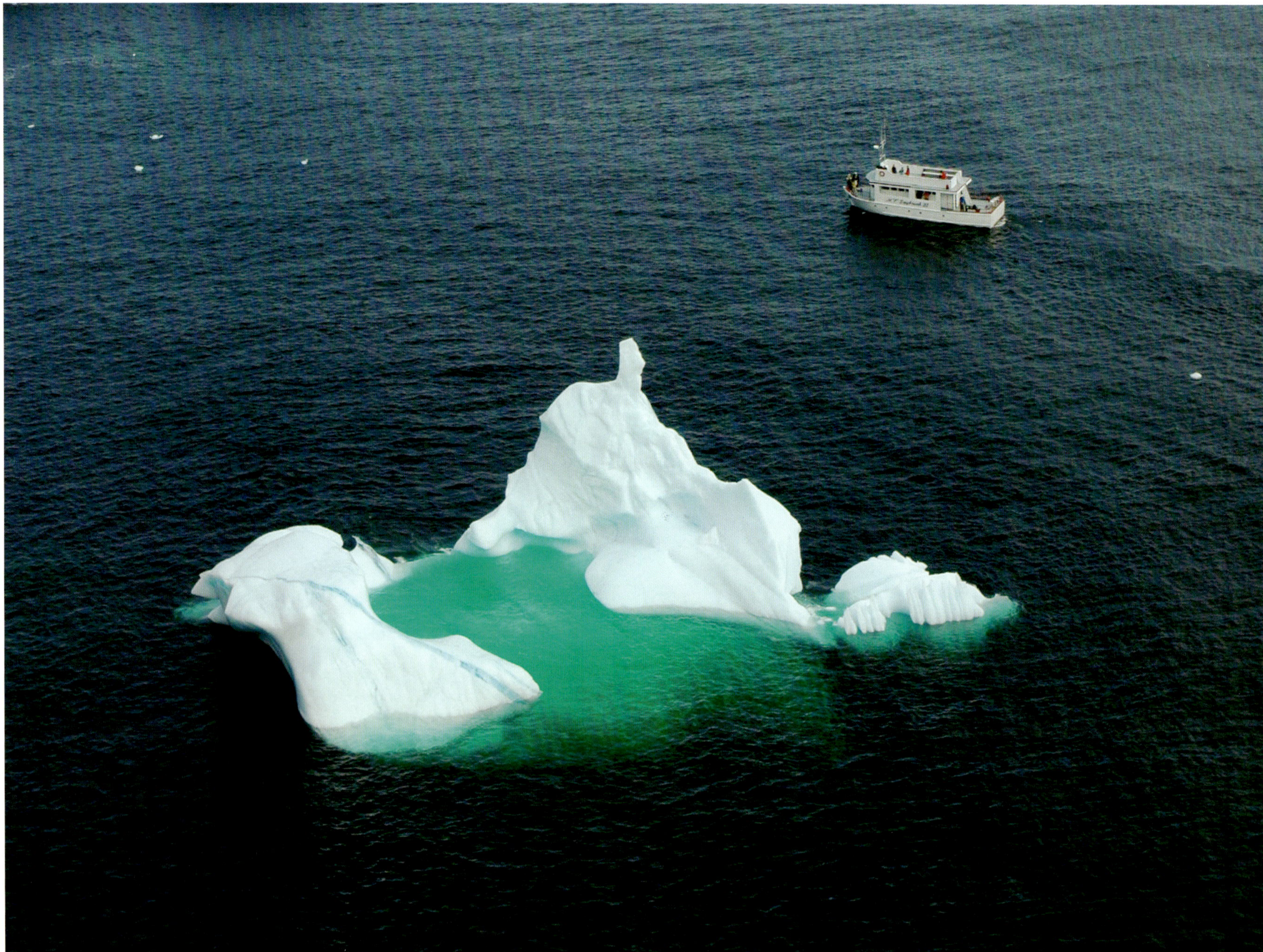

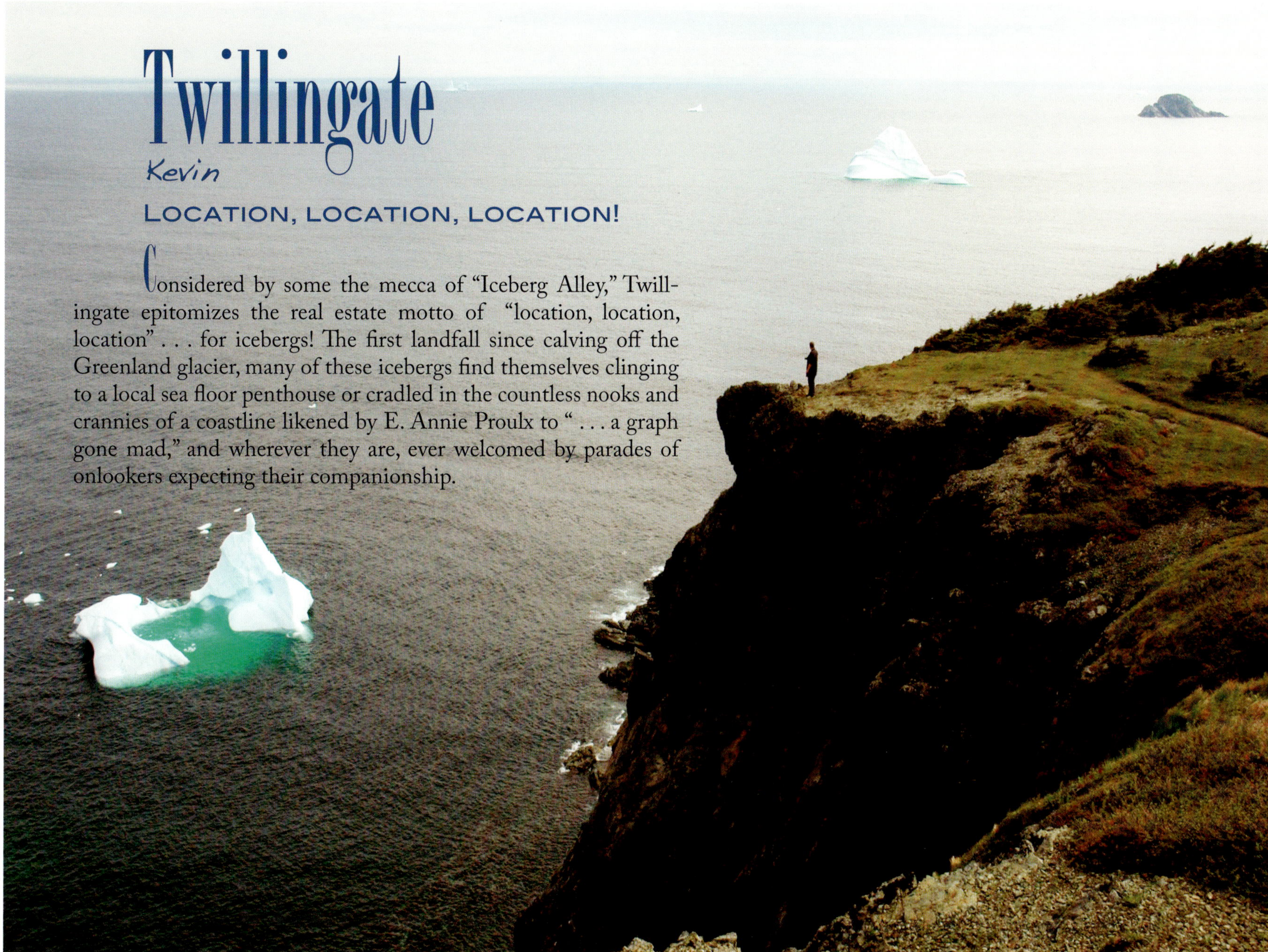

Twillingate

Kevin

LOCATION, LOCATION, LOCATION!

Considered by some the mecca of "Iceberg Alley," Twillingate epitomizes the real estate motto of "location, location, location" . . . for icebergs! The first landfall since calving off the Greenland glacier, many of these icebergs find themselves clinging to a local sea floor penthouse or cradled in the countless nooks and crannies of a coastline likened by E. Annie Proulx to " . . . a graph gone mad," and wherever they are, ever welcomed by parades of onlookers expecting their companionship.

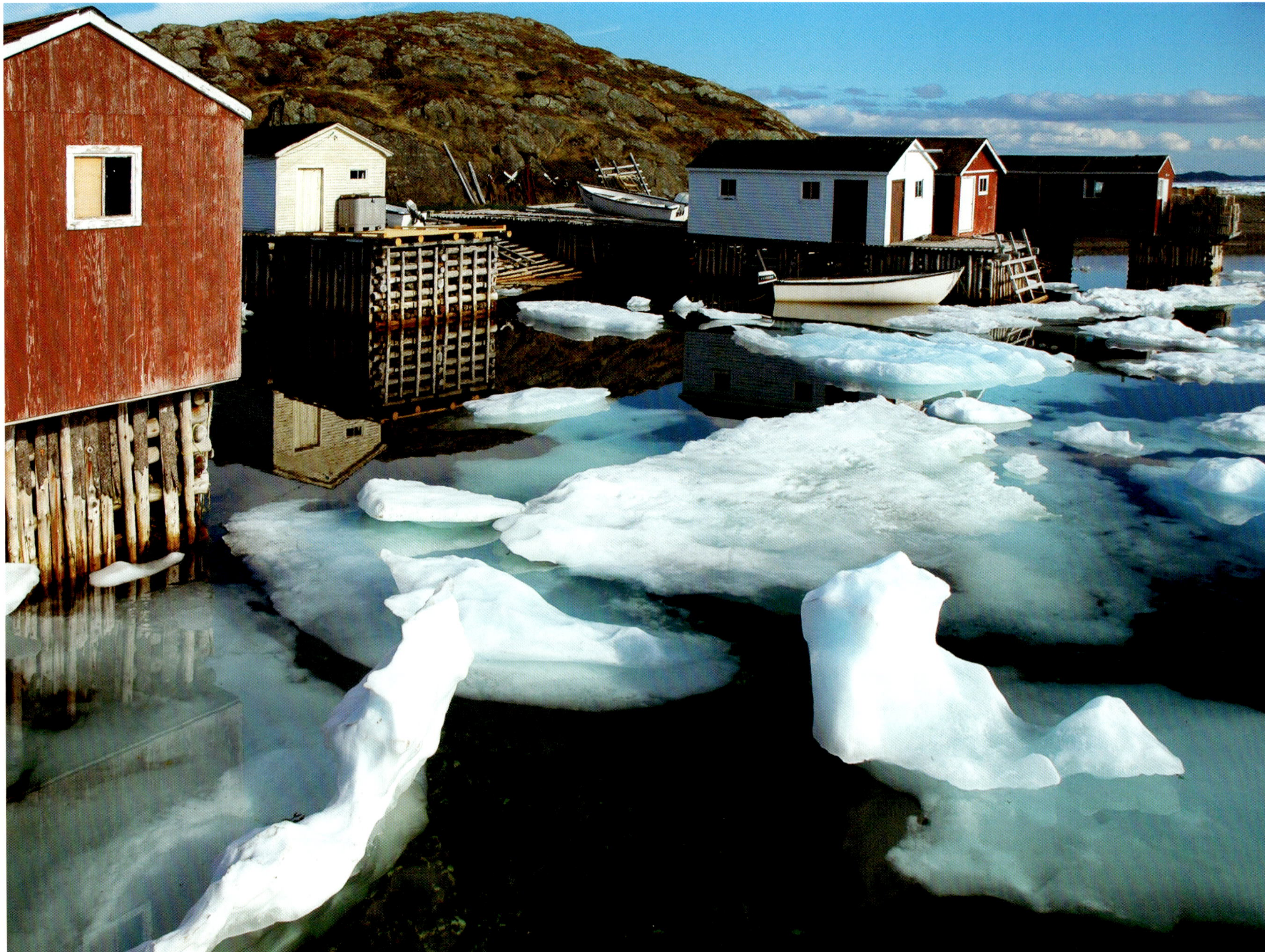

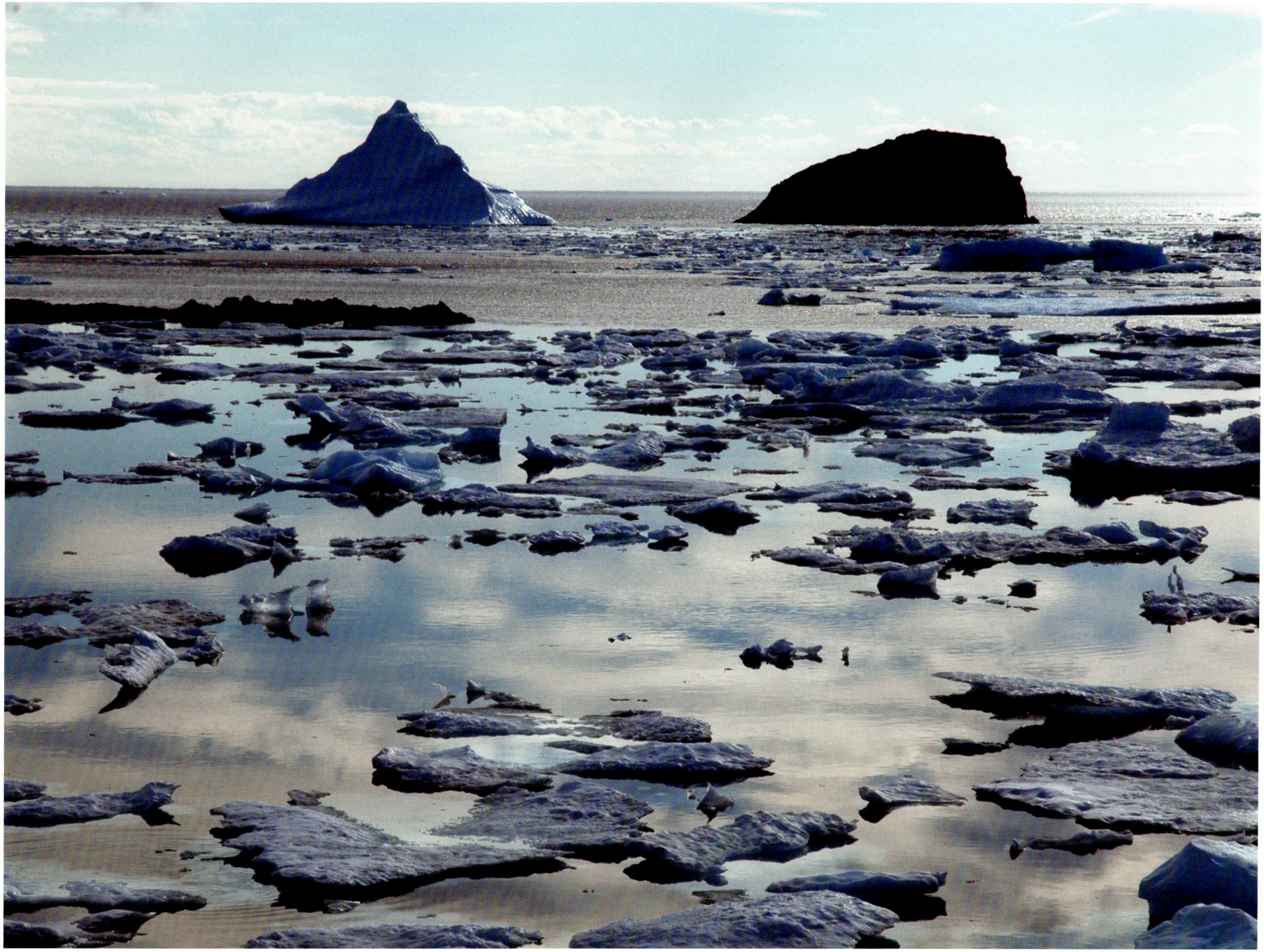

Weapons of Mass Reduction

Steve

I spent two weeks around Twillingate doing iceberg experiments with my colleagues from C-CORE and Coast Guard. Our primary purpose was to test innovative ways of reducing the threat of an iceberg striking a subsea pipeline. Preventing it from getting to the place of concern is one approach, and another is making sure that if a berg does reach the pipeline location it will not strike it. Conventional iceberg management for the offshore oil and gas industry doesn't involve attempts to break up a berg as each part may prove to be a threat equal to or greater than the original whole. However, when one is interested in decreasing the depth that a berg reaches, then it makes a good case for the development and testing of "weapons of mass reduction." Which brings me back to Twillingate . . .

Amongst my favourite memories of that expedition was the shock and alarm of a successful breaking of a small berg – whilst in a speedboat fairly close to the action. The video sequence I recall taking from amidships was informative up to the point where a bus-size block dropped within feet of our gunwales. From that point, the video was limited to the blurry floor of the boat, but what compensated for this break in the data sequence was a fantastic, earnest and unambiguous audio recording of the surprise event. Safe to say that what started out as G entertainment shot into A or R territory, depending on one's interpretation of the oaths.

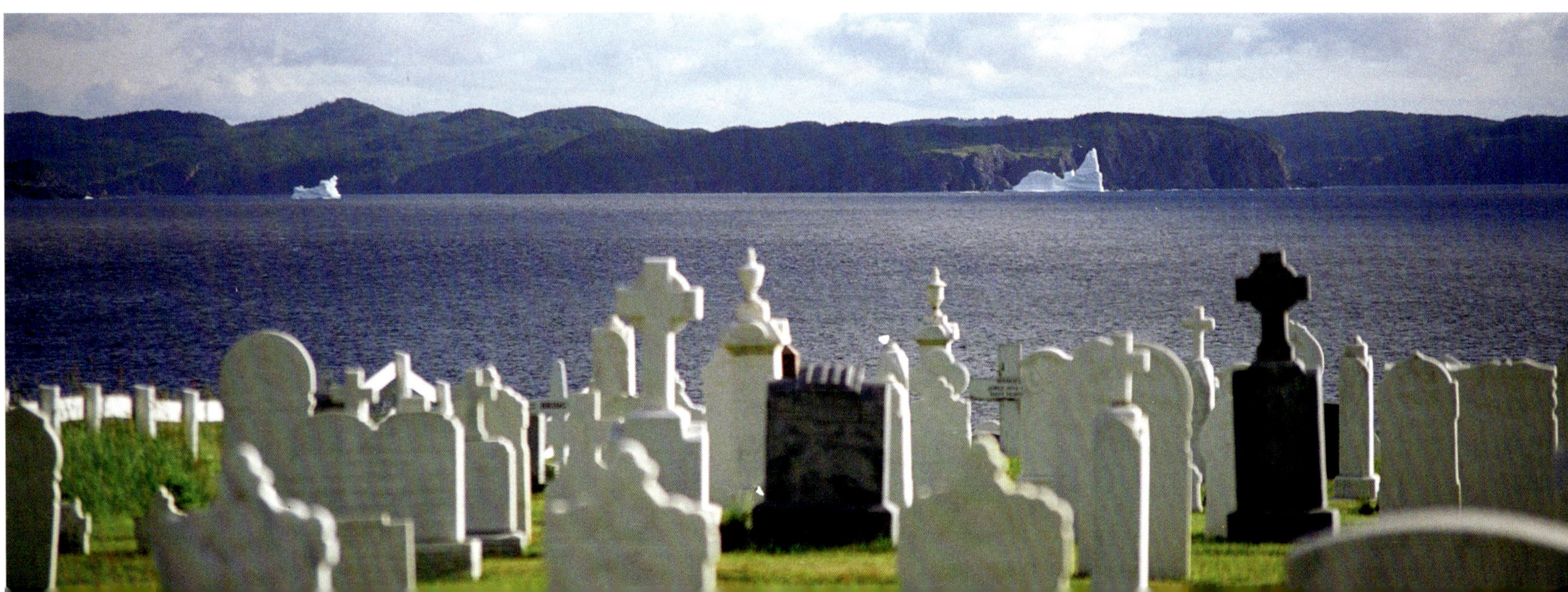

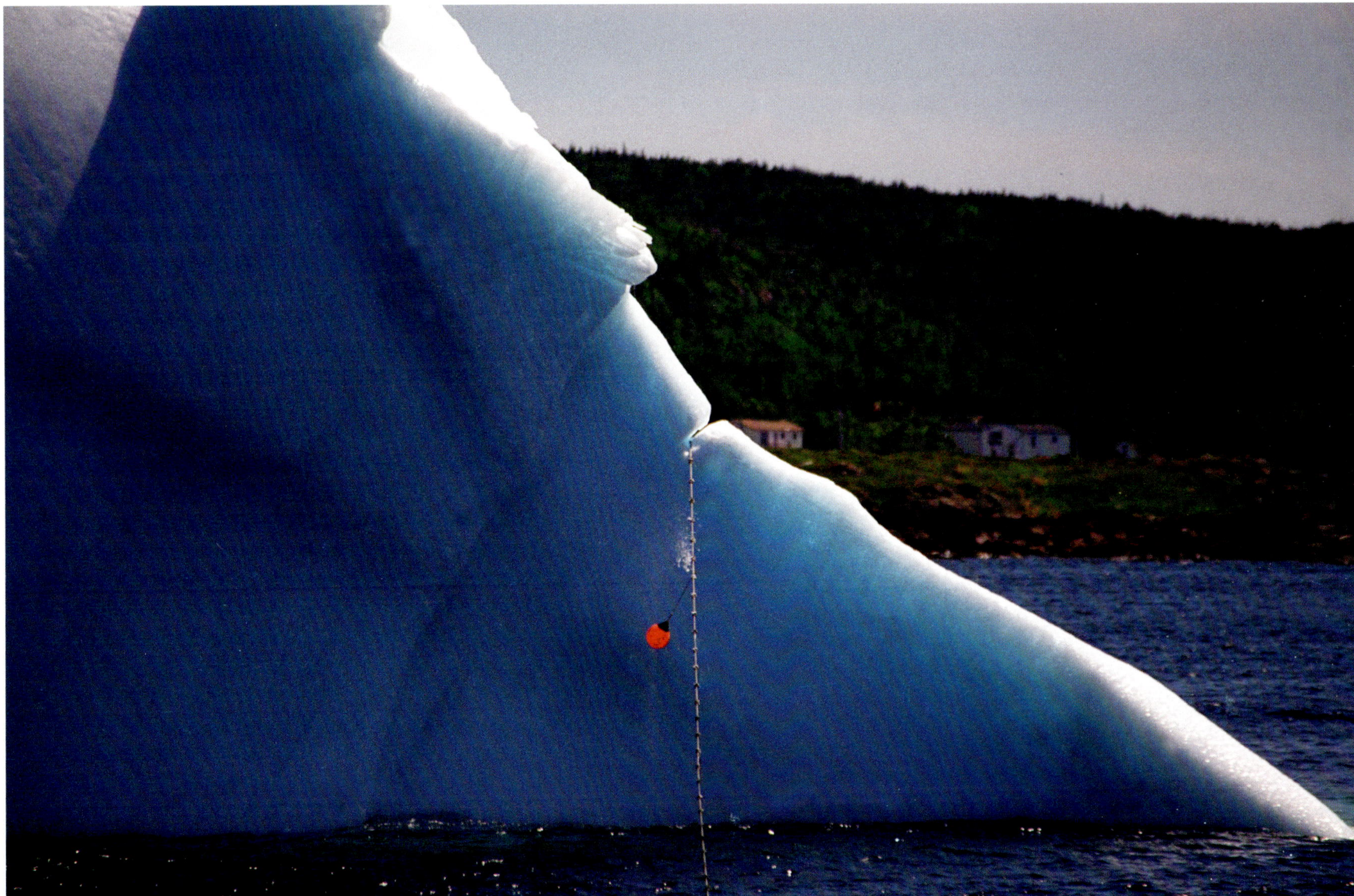

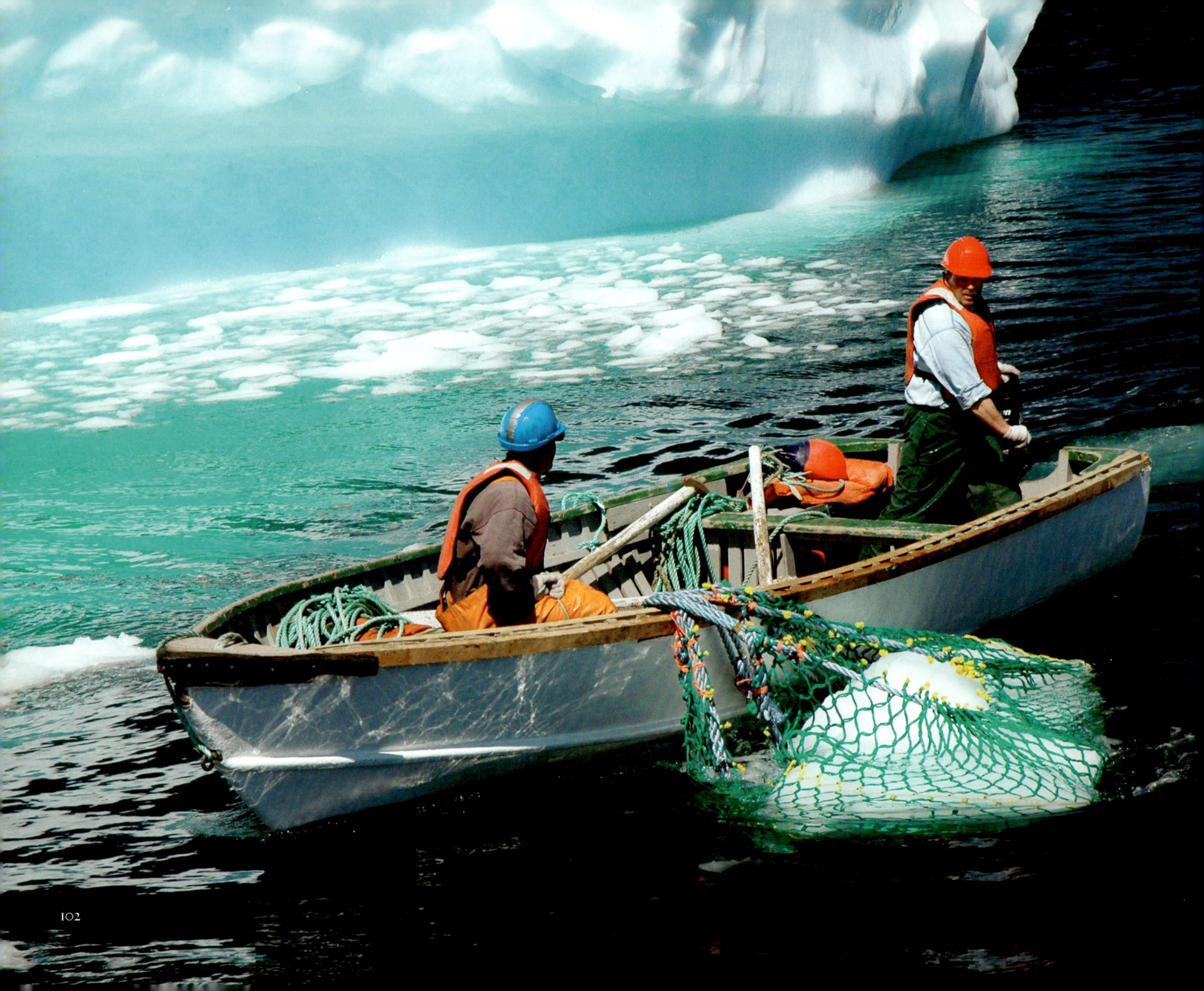

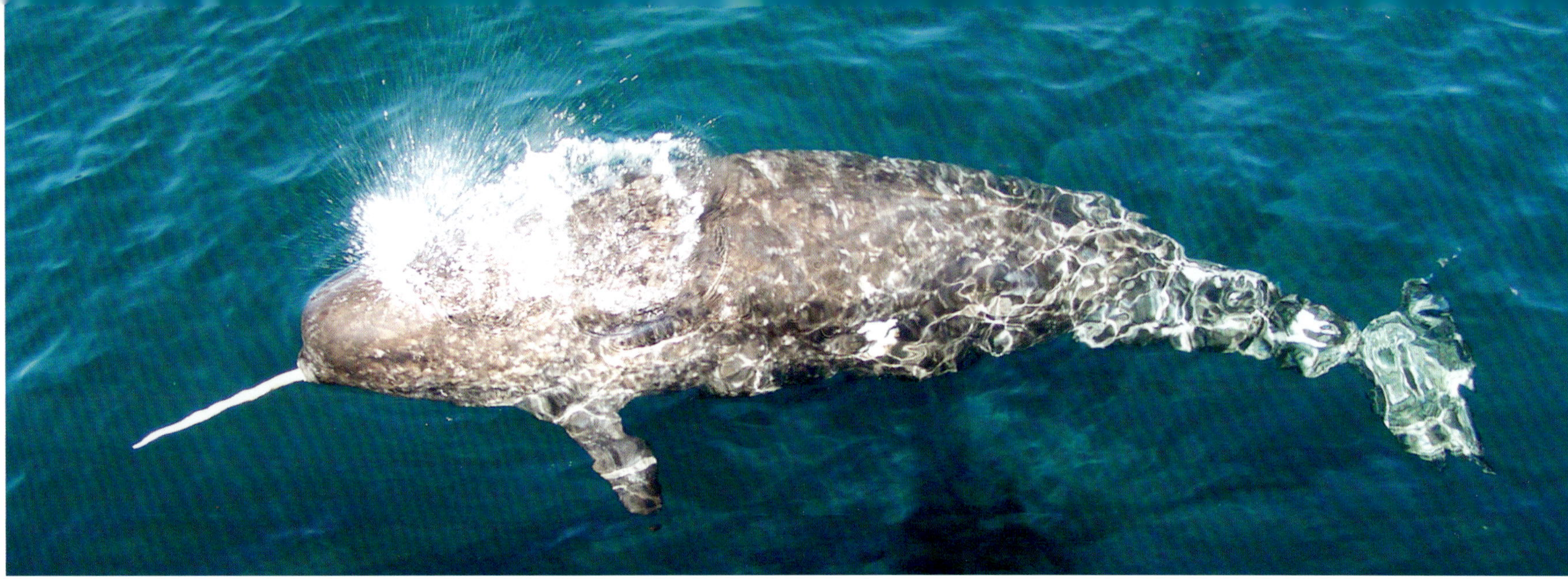

An Ice Ditty

Steve & Kevin

The Ice Collectors

There once was a crew
Out of many they were but a few
Their job of collecting iceberg ice
Was as predictable as a game of dice

Painful was their wait
Their boat *Mottak* trolled a slow gait
Wishing the iceberg to break apart
It was then, their real work would start

To the bergy bits in motorboat they would ride
Around a chunk of ice the net, it was tied
One to hold and one to drive to tow the prize back
Where in the hole or on deck it was added to the stack

Of all the technology tried
Much of which quickly died
Simple was found to be best
A boat of ice always meant rest

On a trip to the "ice fishing" grounds in 2003 Ed Kean had to be careful while operating around a too-curious little narwhale. He said there was an arctic fox on another berg. Seems Ed and his crew have had to work around all sorts of interesting challenges.

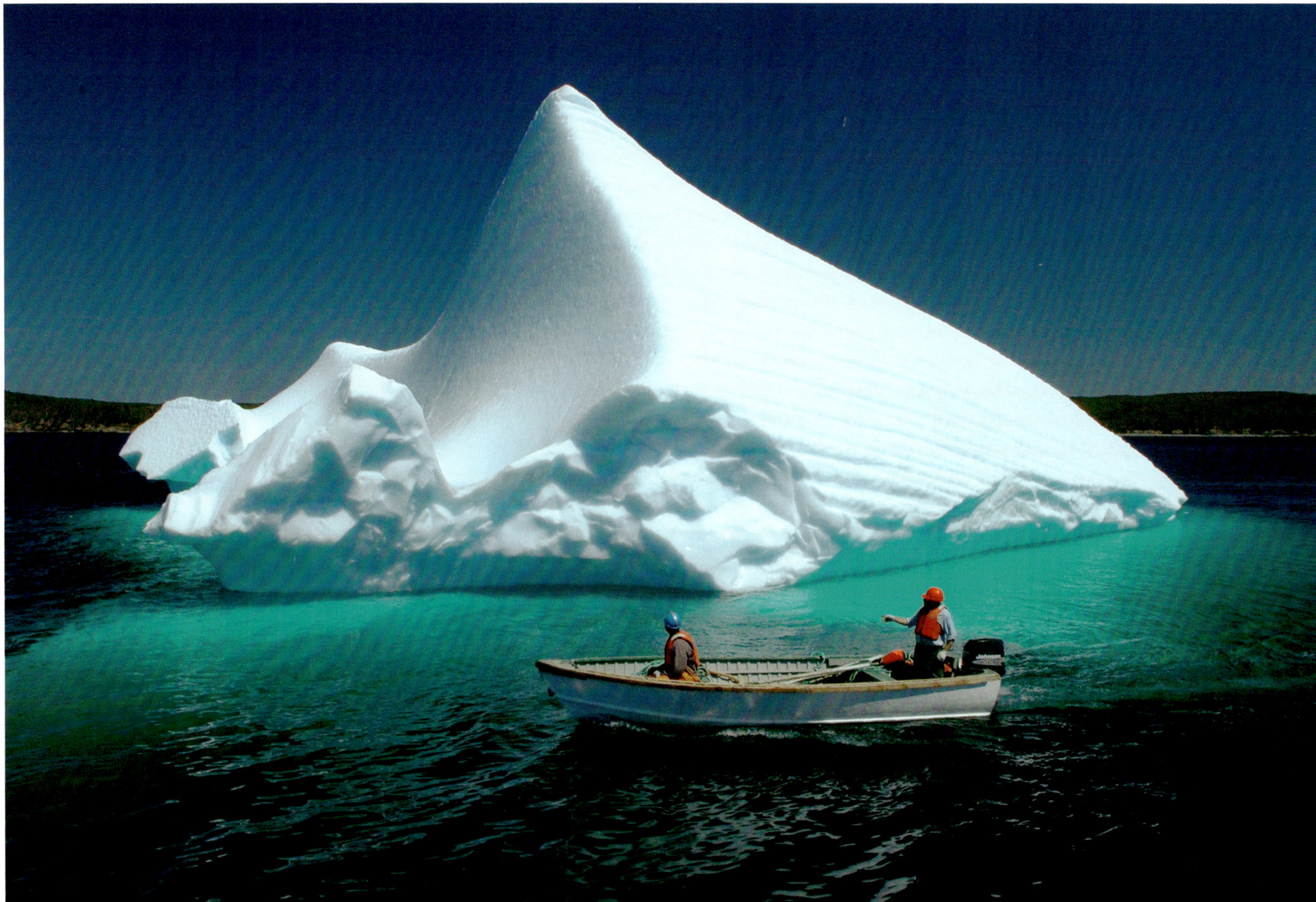

On the Rocks

Iceberg Consumables in Newfoundland and Labrador

Though iceberg ice has been used for generations as a convenient way for watering ships, packing fish and keeping favourite drinks cool, the first mass-produced product to achieve market success is the now-famous ICEBERG Vodka.

The exceptionally smooth vodka is produced by blending triple-distilled alcohol with melted iceberg ice and subjecting these ingredients to several levels of filtration. The final product is bottled in a state-of-the-art facility in St. John's and shipped to customers worldwide.

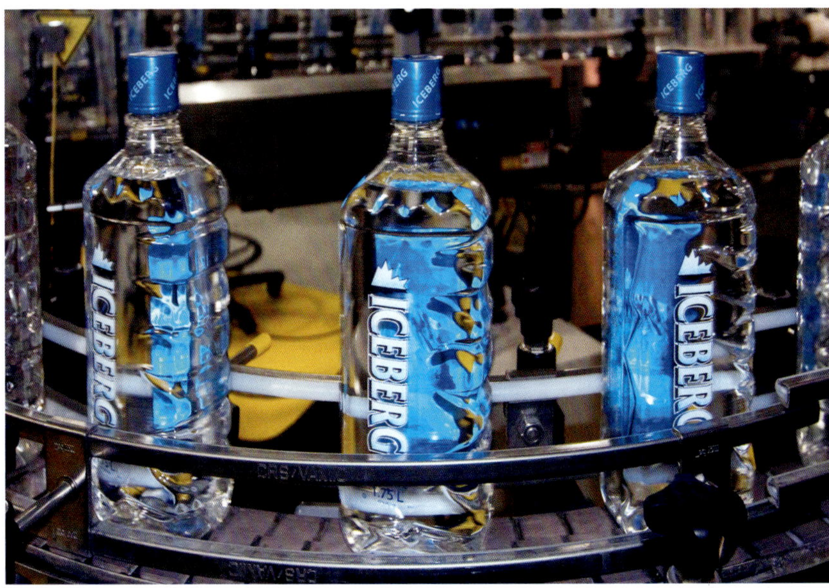

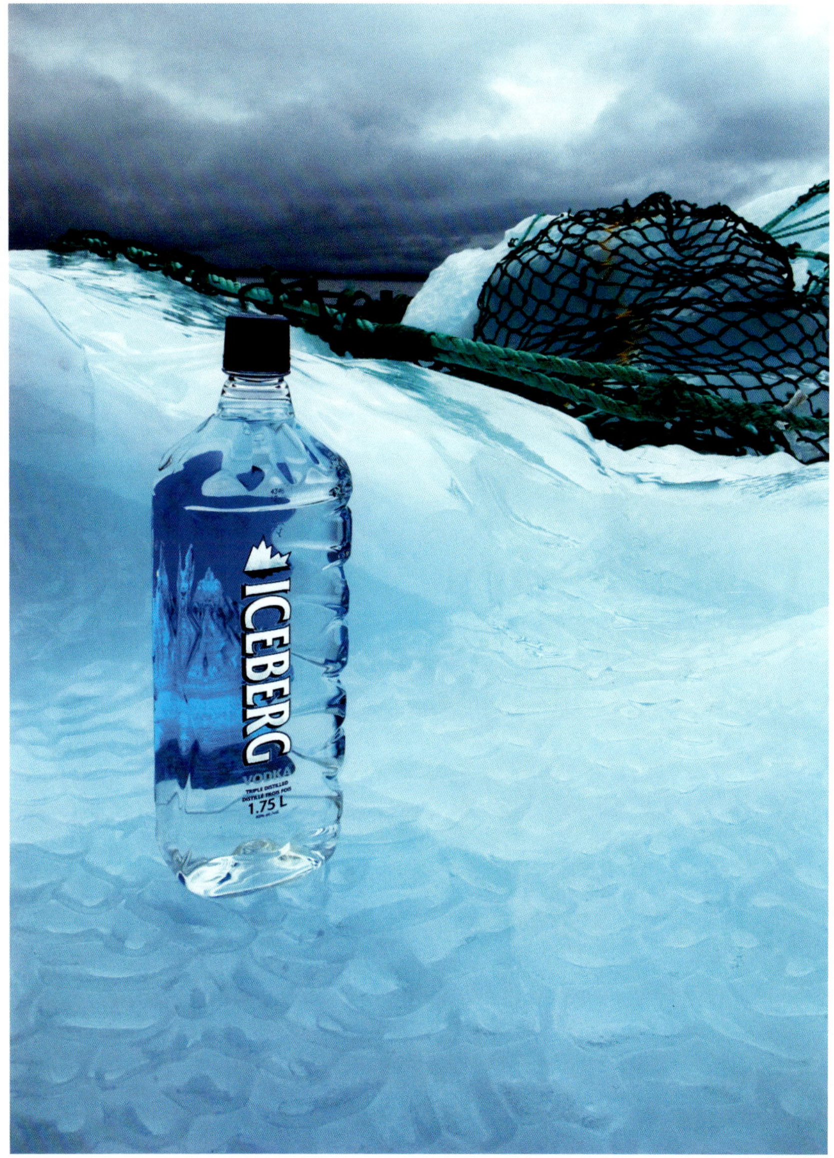

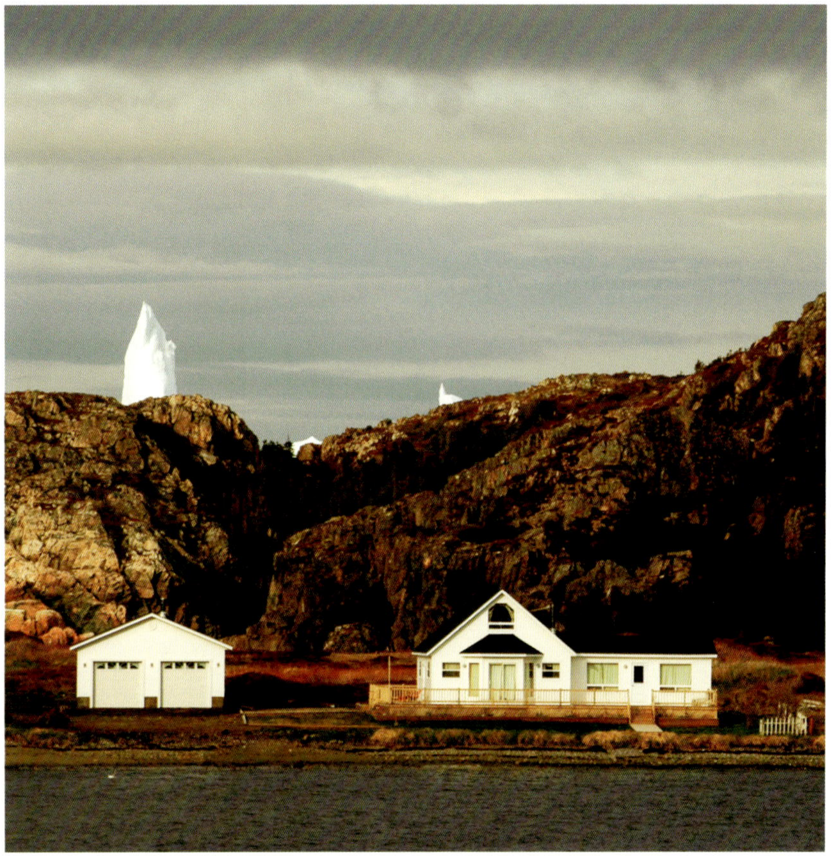

Towers and Turrets

Kevin

ROUND HARBOUR, MAY 2008

The twin turrets of ice towering over the distant hillside were enough reason for us to shut the car off. We discovered that driving a 1994 Ford Mustang Shelby Cobra convertible at twenty to thirty kilometres per hour for an hour looking for icebergs was akin to environmental terrorism . . . we had burnt, what seemed to us at the time, over a half-tank of gas for one iceberg.

En route to the Little Harbour iceberg we passed two concrete root cellars and two building-like homes, one of which had seen better times. A faded sign over the abandoned homestead advertising hamburgers and hot dogs indicated to us that this was once a thriving community.

What we planned as a short forty-to-fifty-minute turnaround trip to the iceberg turned into a four-hour pilgrimage. Remnants of a recent fracturing coated the water's surface between the berg and the coastline. No less than four rams of ice spiralled skyward, all attached to a common base. As we departed we agreed, this was an iceberg that with time was sure to display drama!

Looking at the gas gauge as the car started for the homeward leg of our journey we surmised, "Some cars are meant to drive like stink!"

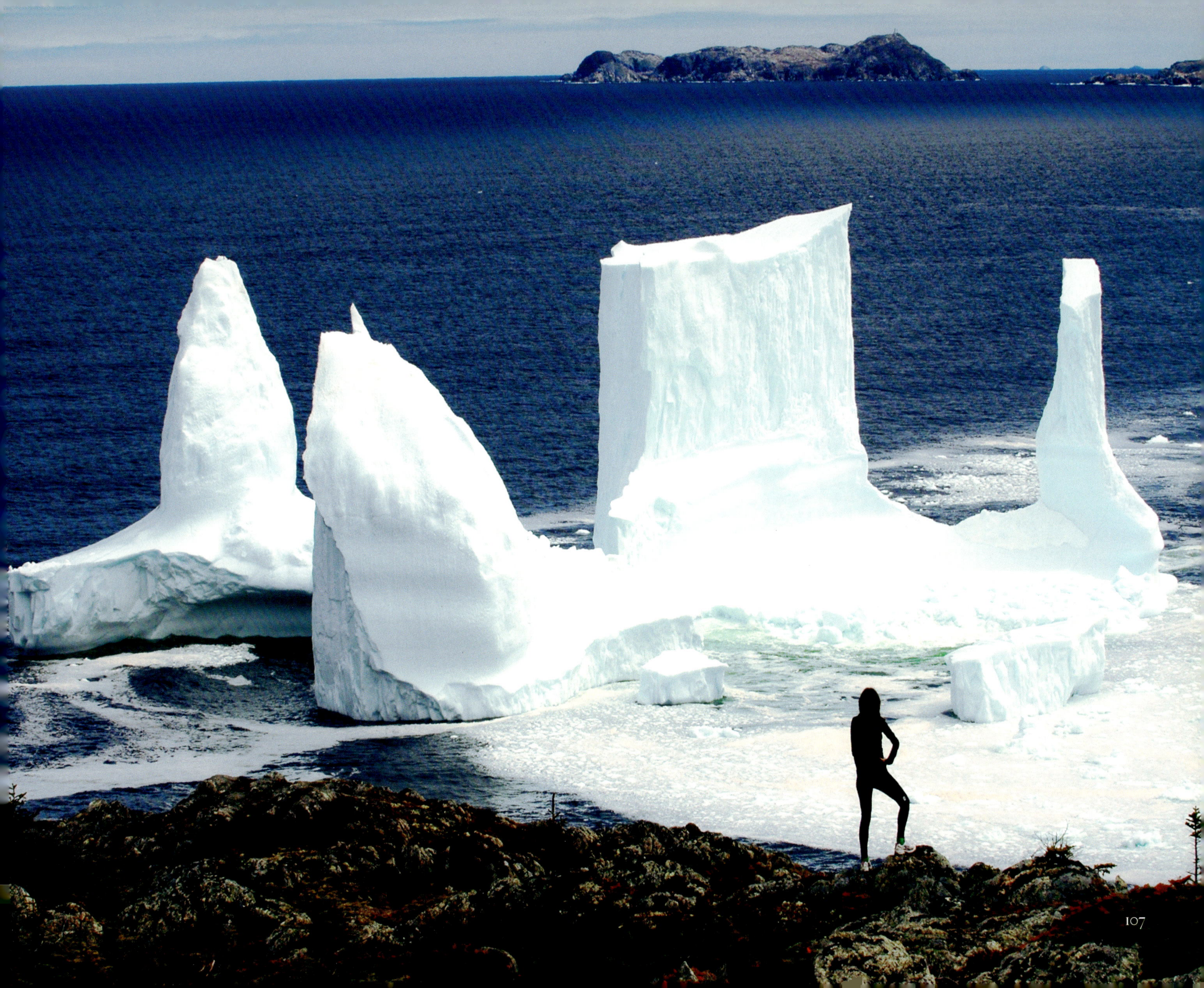

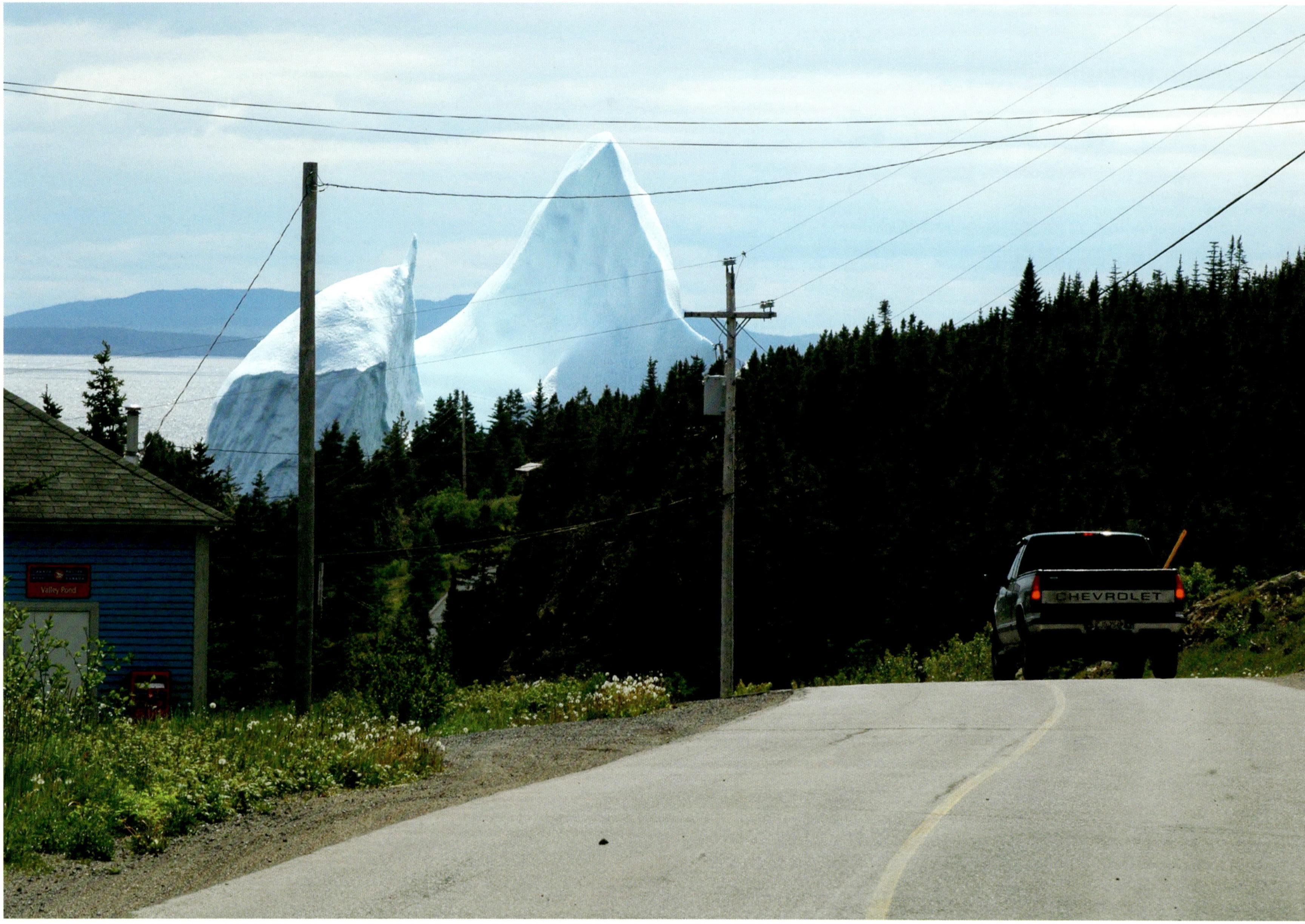
Valley Pond
CHEVROLET

Pilgrimage

Kevin

Many holy shrines are located close to nature, on mountain peaks, riverbanks, or seashores, where the setting is conducive for meditation and reflection. In many rural communities of Newfoundland and Labrador, the church was built on a hilltop, its steeple rising above all other man-made structures in the community, likening this edifice as your clearest avenue to the "One" above.

The "shrine" of Valley Pond in 2007 was a towering iceberg that despite being some hundreds of metres offshore soared skyward above the town. "Pilgrims" from near and afar gathered atop the barren seaside knoll. The light breeze, like a telephone party line, made it impossible not to discern stories being told nearby. Local folks revealing past loves and romantic times along these majestic shores, tourists thousands of kilometres from home marvelling over the spectacle of ice massaged by the wind, moon, stars, and sun for some 10,000 years. Today, caressed by the water, wind, and sun, the iceberg reveals its dying cries with periodic thunderous booms, precursors to losing another piece of itself and symbolic of both its strength and fragility.

How ironic, I thought, that it is something from Greenland bringing together this mosaic of cultures on this remote, isolated coast of this earth! "Religion points to that area of human experience where in one way or another man comes upon mystery as a summons to pilgrimage." (Frederick Buechner)

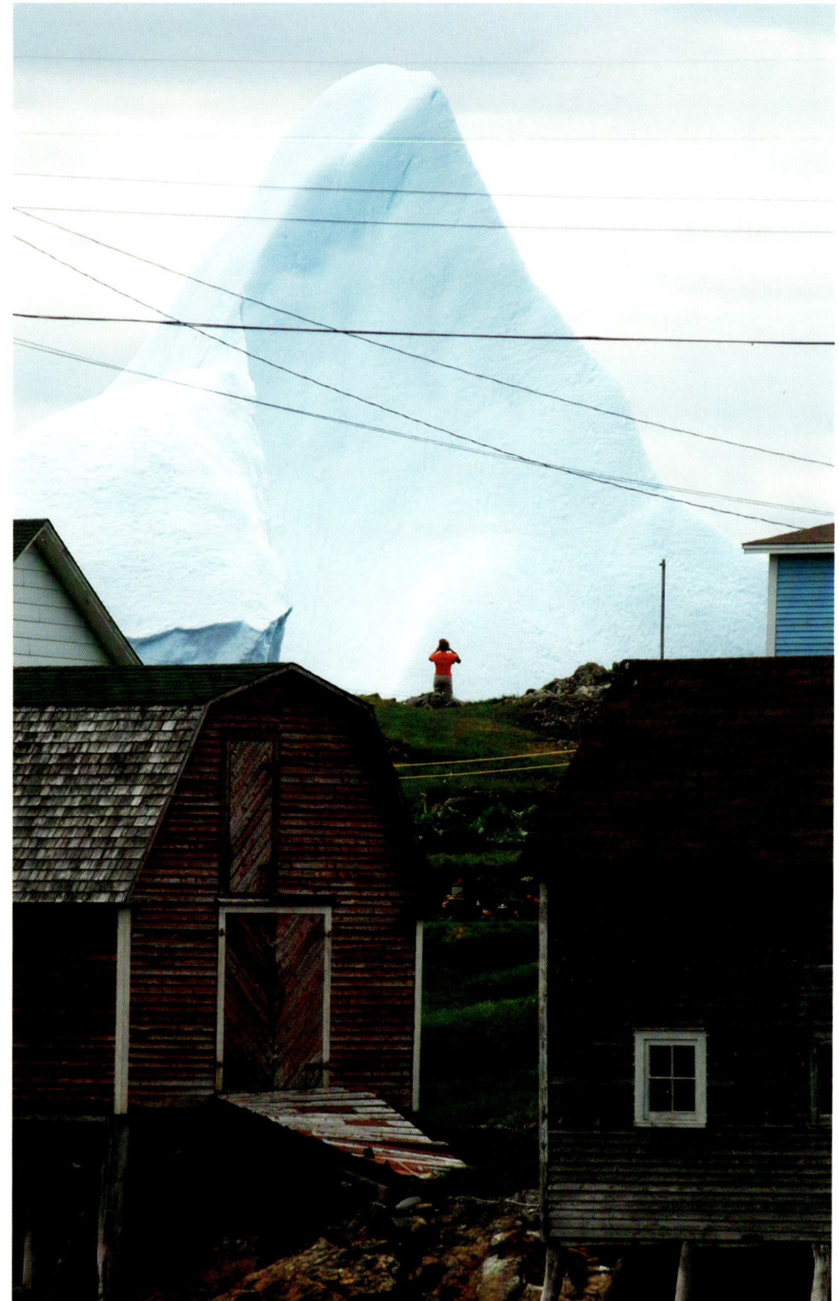

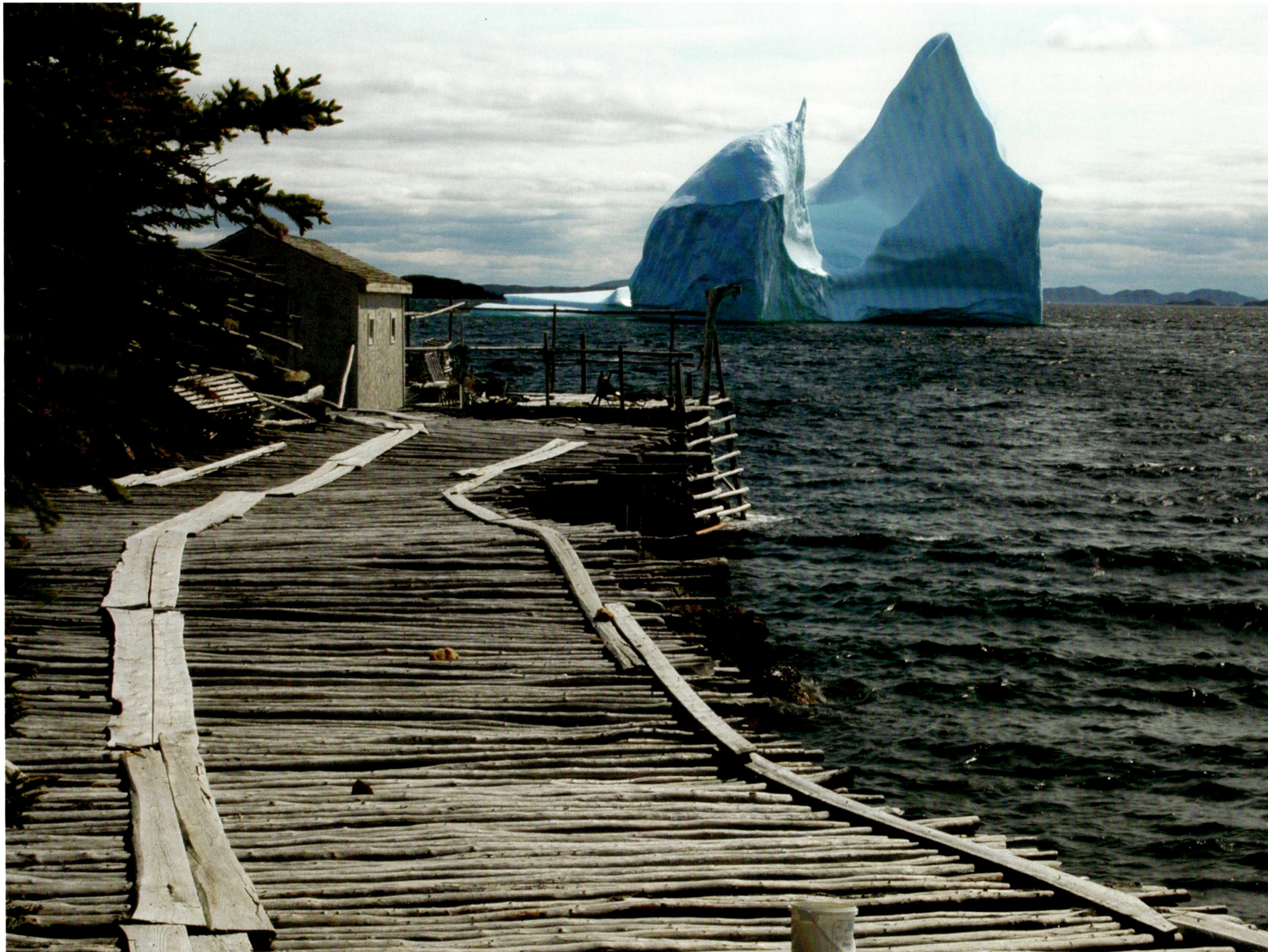

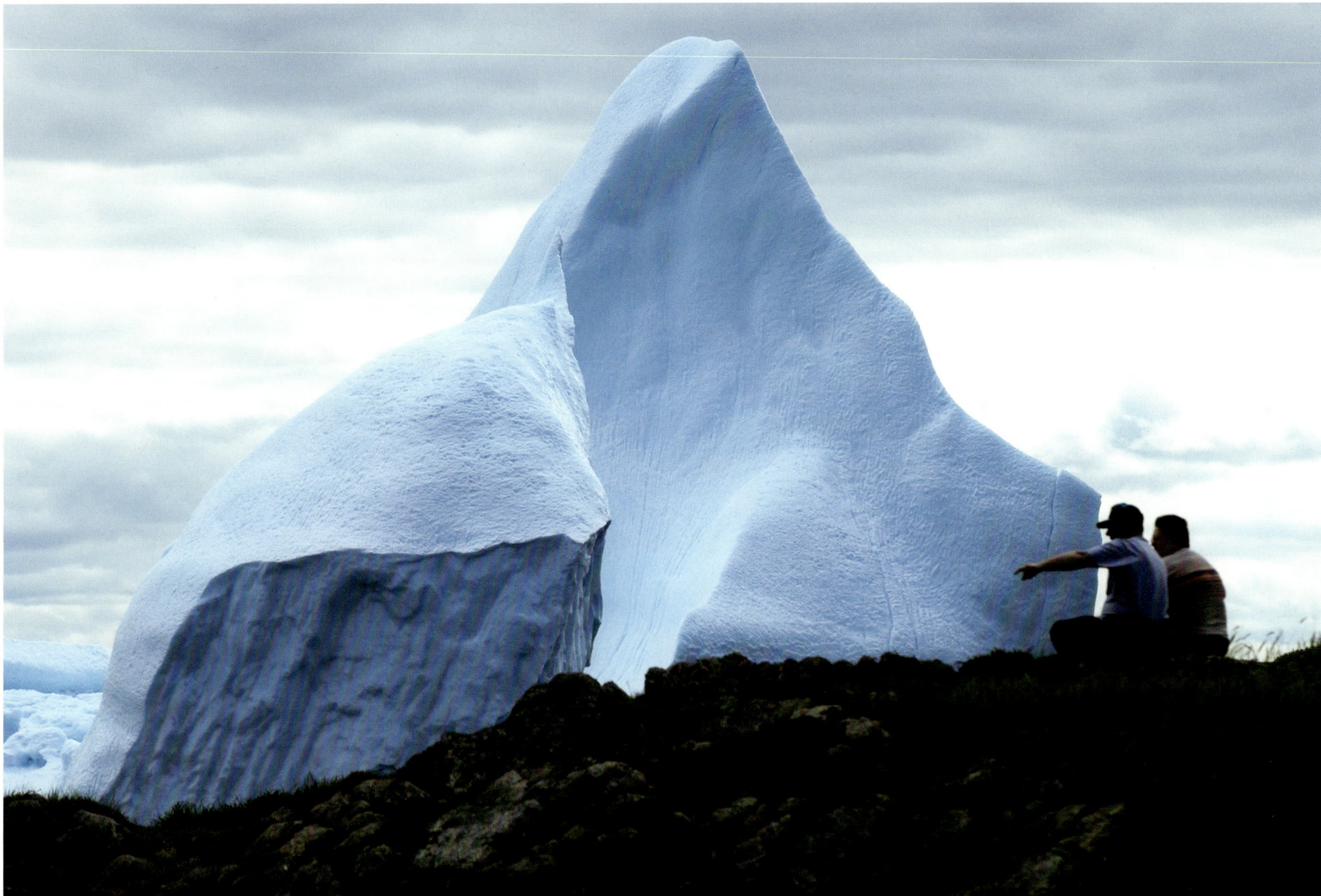

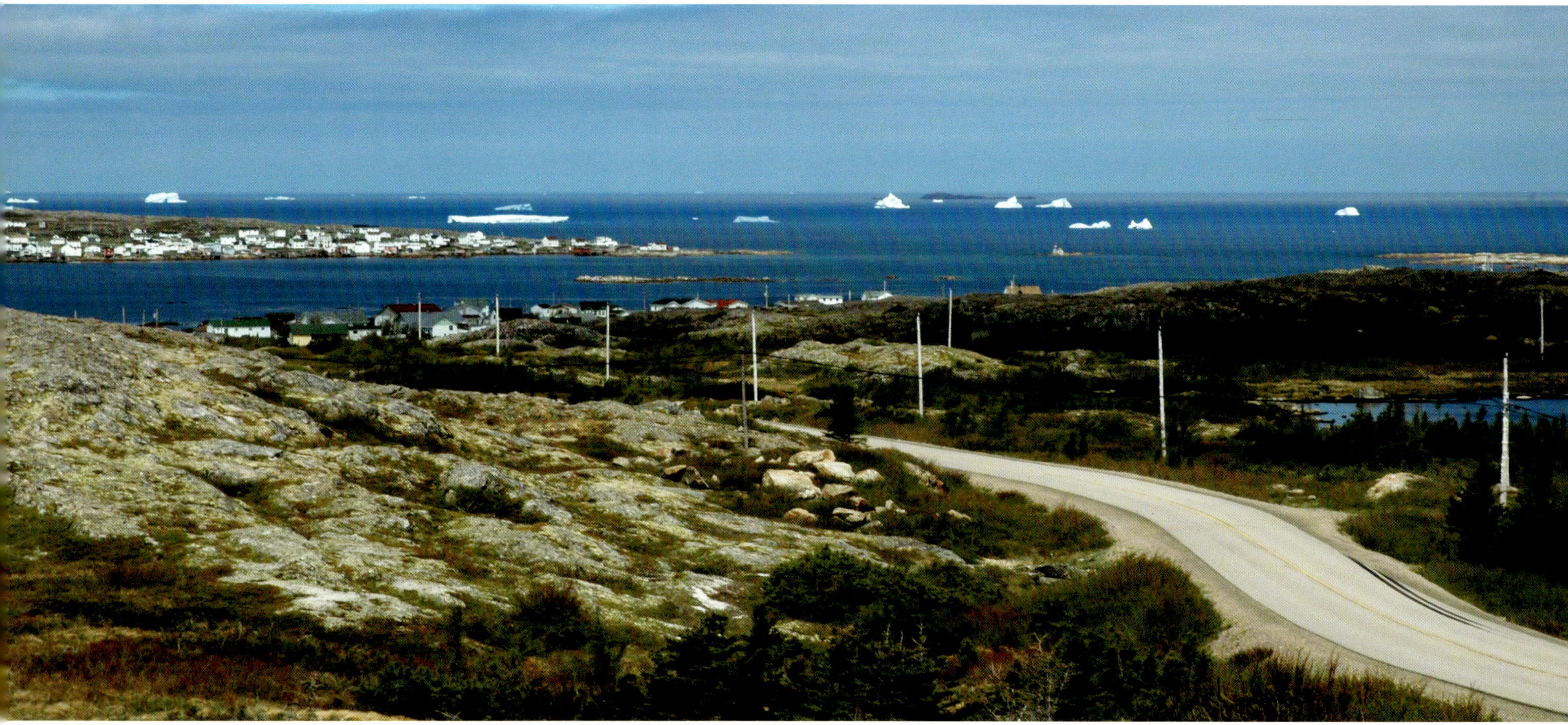

Never the Like

Kevin

Northeast Coast: Joe Batt's Arm

In my search for icebergs it was during a call to Cyril McGrath of Tilting on Fogo Island that I learned the meaning of the colloquial expression "Never the like!" On asking Cyril, "Any icebergs out your way?" he replied: "Never the like, never the like . . . I haven't seen this many icebergs in fifty years!" Off of Joe Batt's Arm (pronounced j'btts'arm) 'tis a job to see the water for the ice!"

The community of Joe Batt's Arm envelops a crescent-shaped cove. To the north it is caressed by the cool waters of the North Atlantic Ocean's Labrador Current. To the east, south, and west of the community is a barren rocky moonscape that appears to defy human habitation. Yet over the past 300 years, its inhabitants not only survived, they thrived making a living from the riches of the sea. Today, the riches of the sea were not fish; it was ice, and there it was aplenty! Yet all the icebergs were a little off shore, the distance diminishing their grandeur.

Today, a light southwesterly breeze, combined with a warm summer sun and relatively cloudless sky made for a "good day on clothes"; the clotheslines in the community were full by 10:00 a.m. Many of these laundry lines made for a nice anchor to the outside ice.

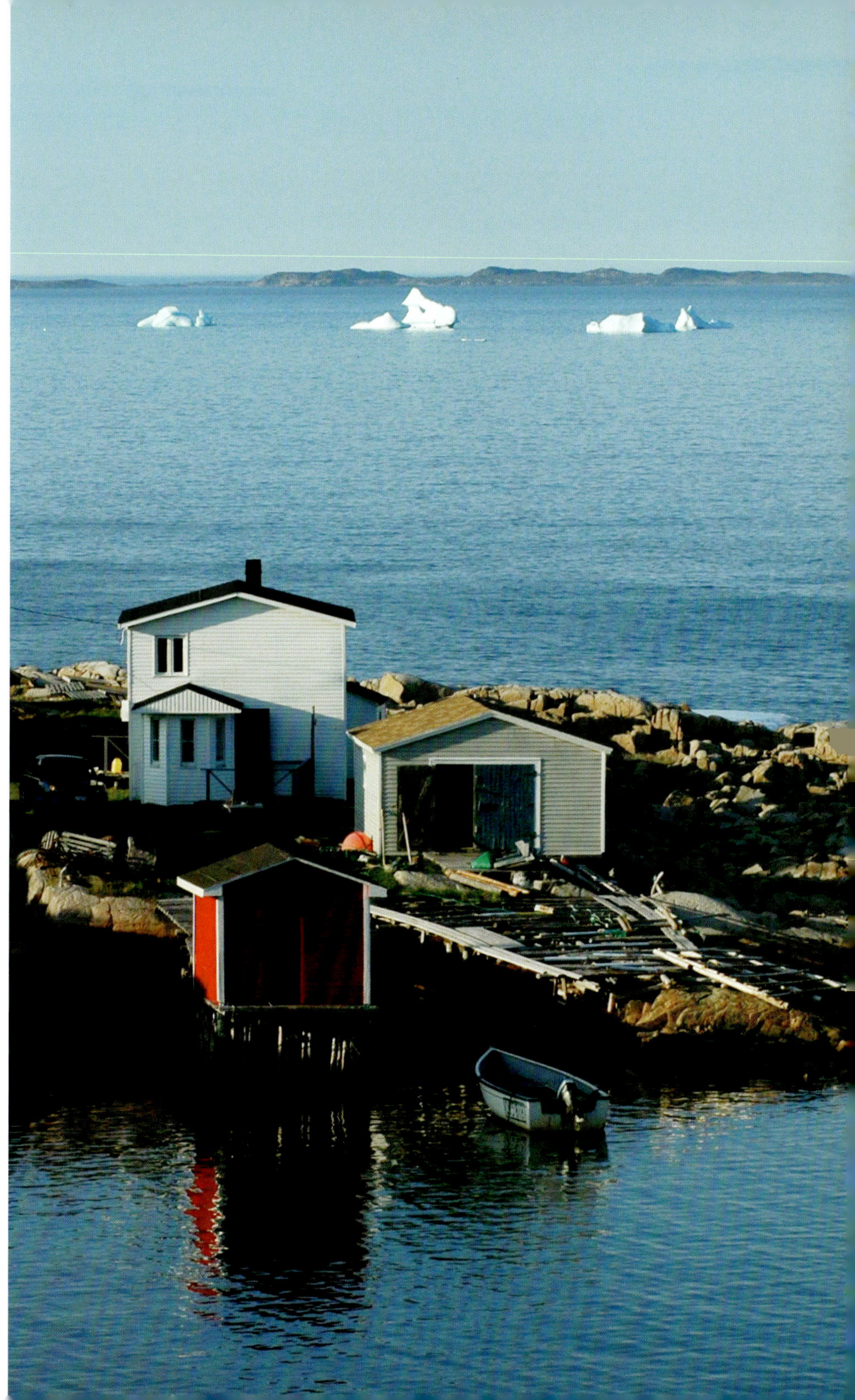

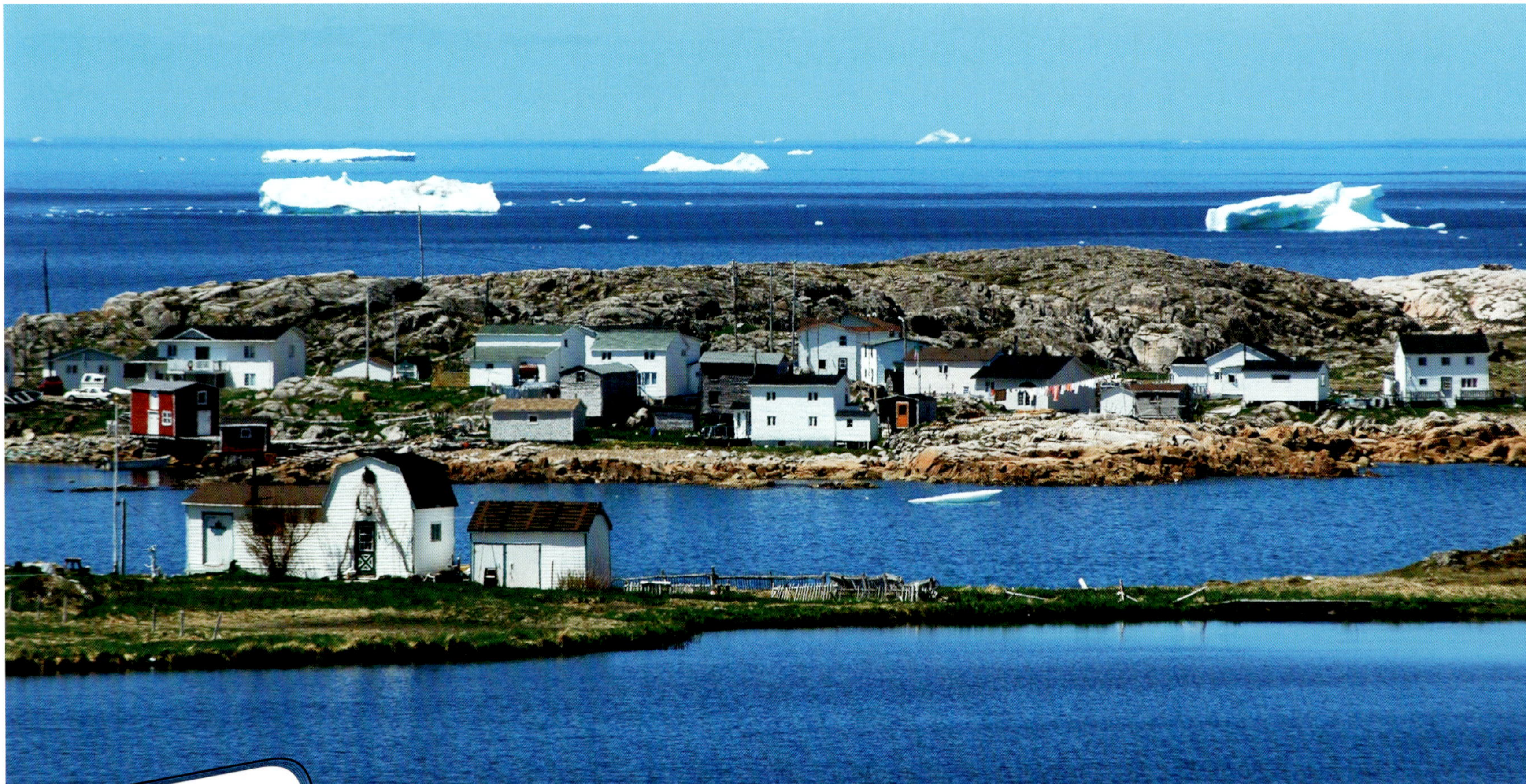

Note to self:

Interesting that back as few as fifty years ago communities were divided along religious lines, with Joe Batt's Arm being Protestant and Tilting being Catholic . . . and this was called the "free world"?

Guarding the cove of the community were two ship-resembling icebergs. To the west floated a giant slab of ice resembling an aircraft carrier while grounded to the east was what appeared like the hull of a Spanish Galleon. Most striking to me was a single iceberg casting two human-like figures engaging one another: a paternal face regarding a shapely debutante in a fitted crinoline gown. The armada of icebergs off Joe Batt's Arm, especially for me, was definitely "Never the like!"

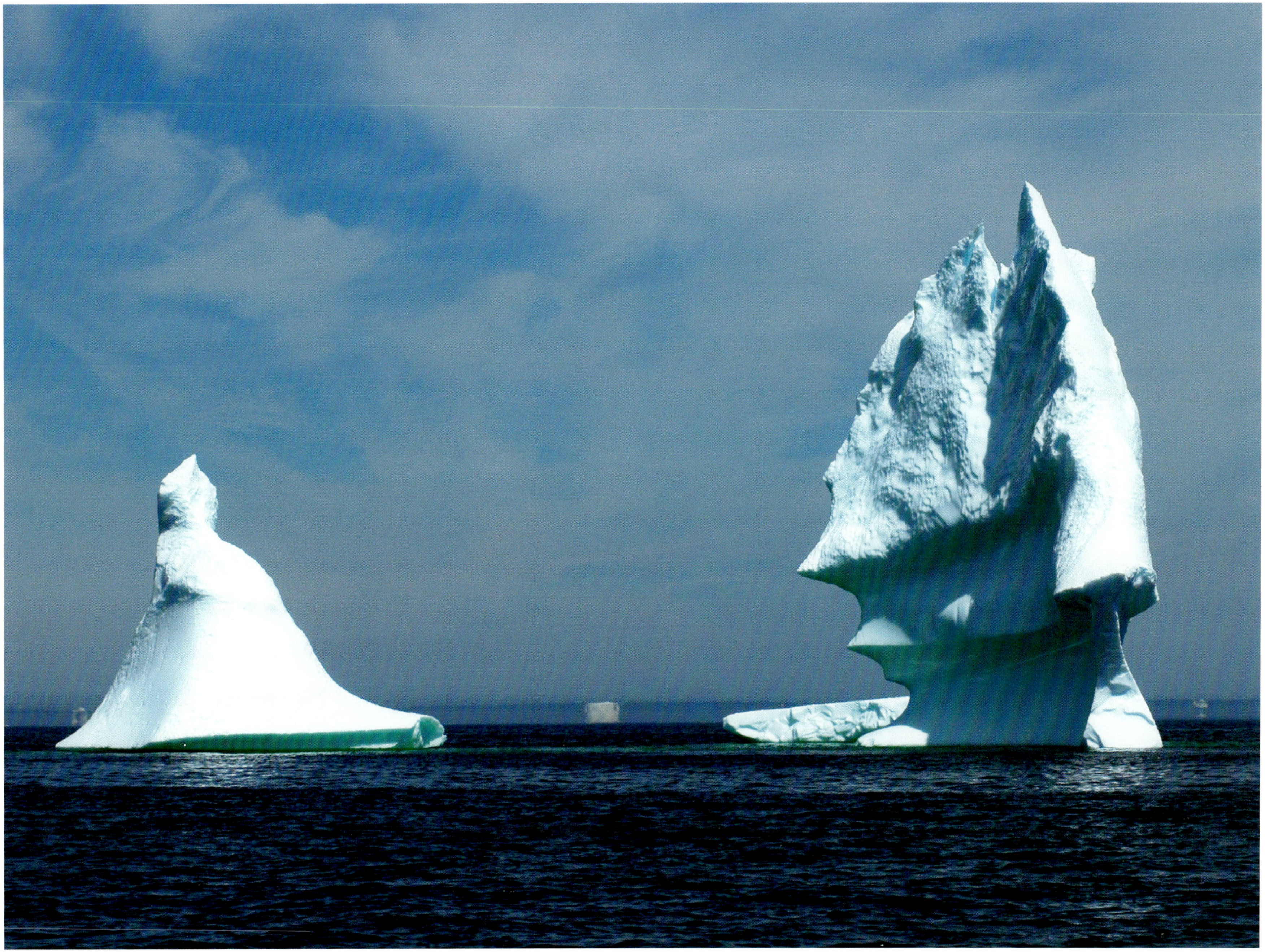

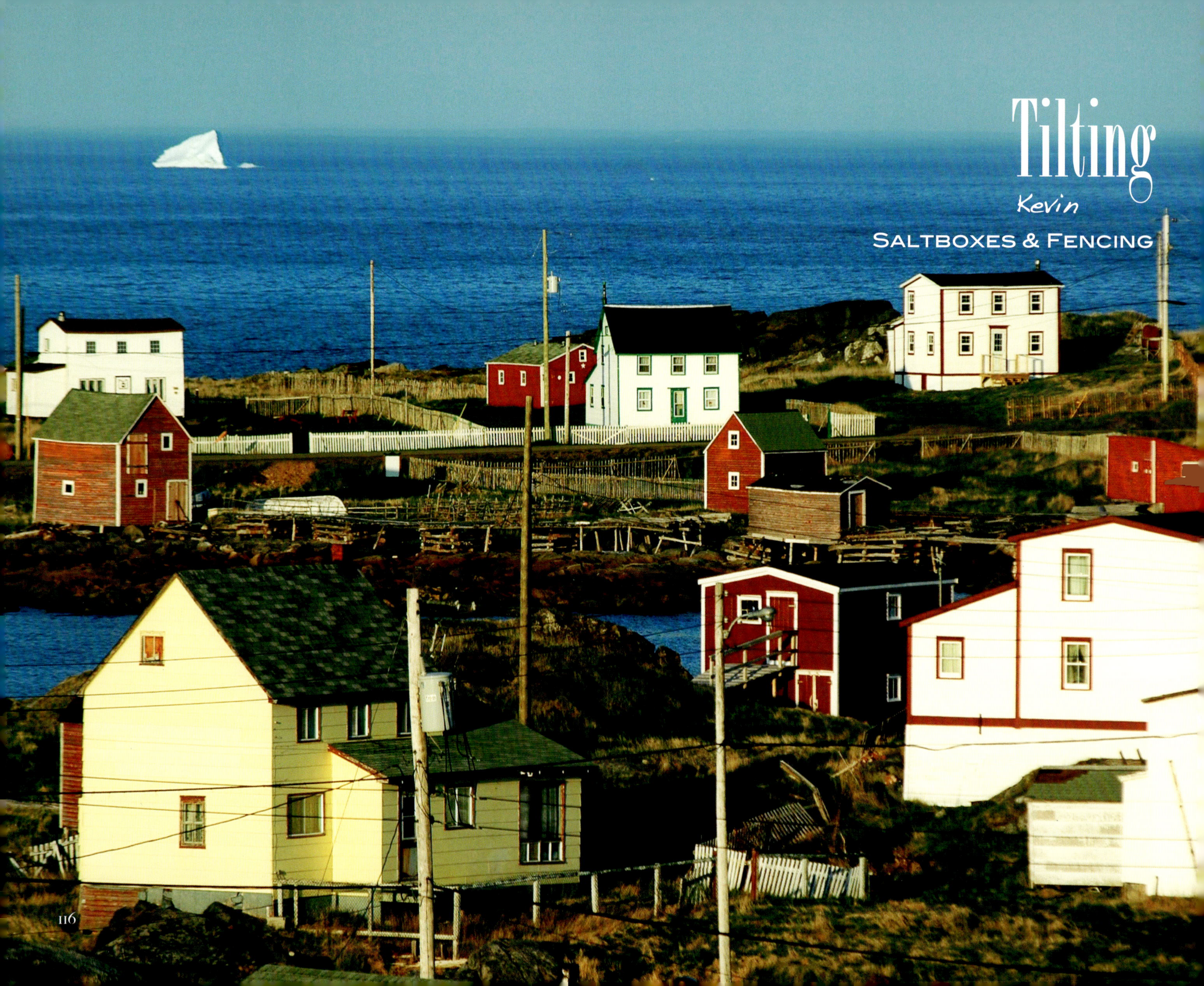

Tilting

Kevin

Saltboxes & Fencing

The stages, stores, saltbox homes, root cellars, gardens, and fencing of Tilting are inspirational. Pristine in rustic and weathered appearance, but clean and tidy like a living museum. Washed with salt spray, scoured by driving rain, cleansed by hurricane winds and renewed with growth from a healthy and lengthy spring. Well maintained in traditional styling, the community of Tilting has been the visual anchor to many an iceberg over the years and the scene in the community today not that much unlike anything in the past one hundred years.

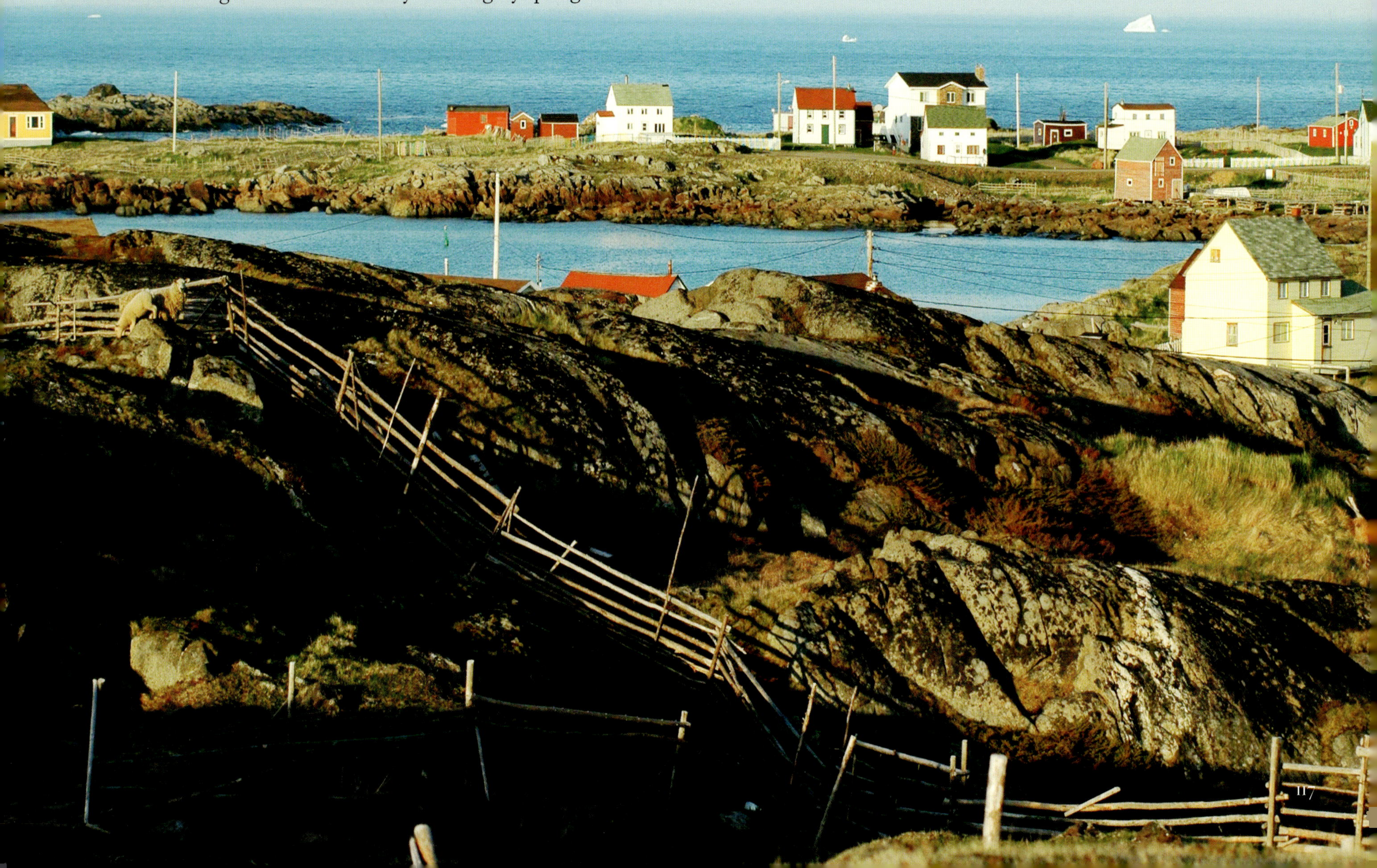

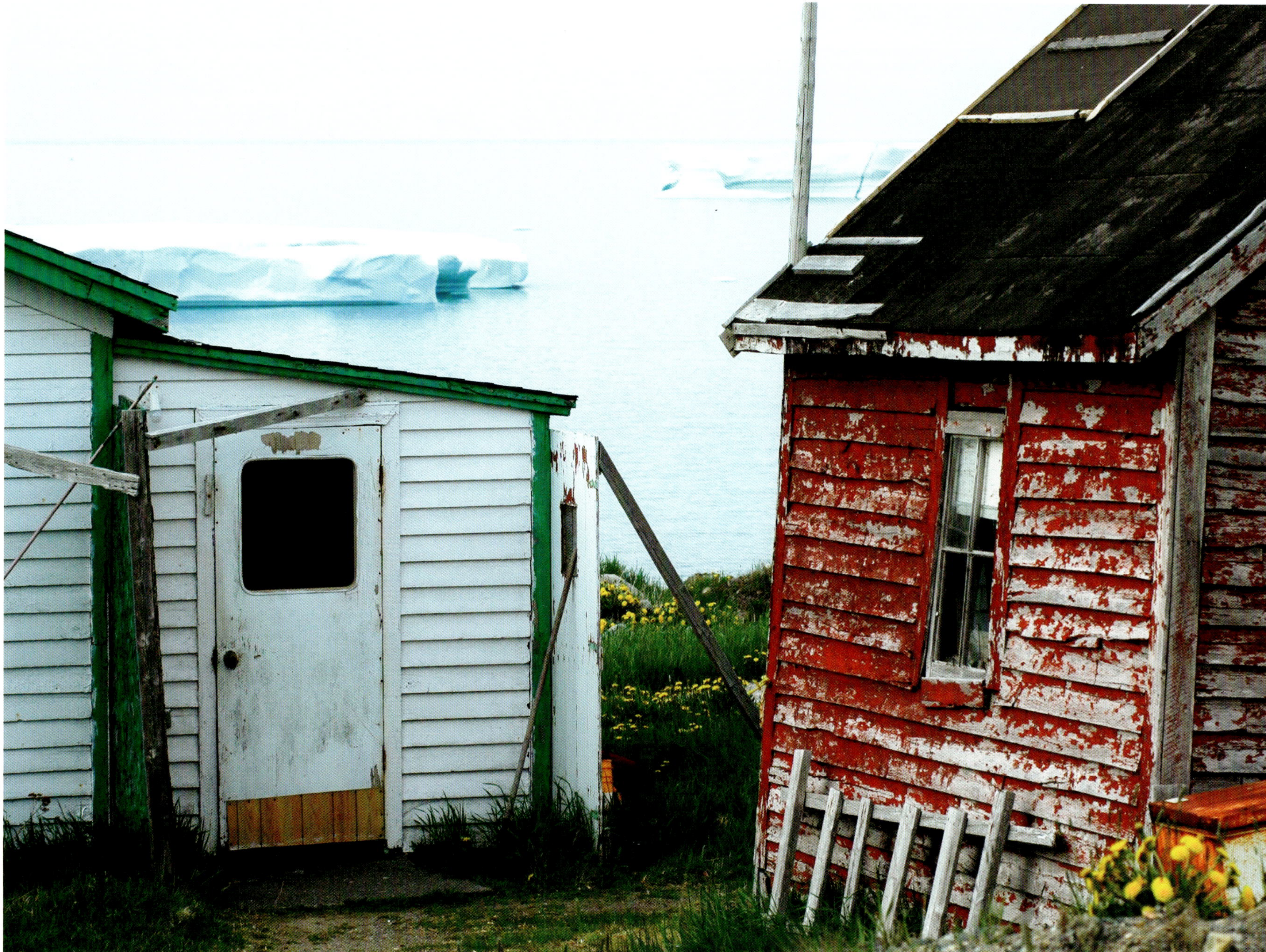

Sunday Best

Kevin

BONAVISTA PENINSULA

Today was an idyllic summer day, where past spring rains fostered lush green valleys, and flowering plants, bathed in today's bright sunshine, blossomed! It was the kind of day, after a long winter and challenging spring, people come out of the woodwork in "droves."

There must have been fifty icebergs along the north shore of the Bonavista Peninsula, making it a destination of choice for many out for a Sunday drive. Convertibles, tops down and hair drifting in the wind; sedans loaded with kids enjoying ice cream cones and icebergs; pickup trucks with enamoured girls wantonly seated in the middle seat; motorbikes rumbling in groups pausing at each cove for another spectacle of nature at its best! Of the hordes of people encountered that day, most striking was a Harley Davidson couple dressed in obviously fitted leather, complimented with matching helmet headsets, and the lady on the back of the bike, wearing four-inch stiletto heels . . .

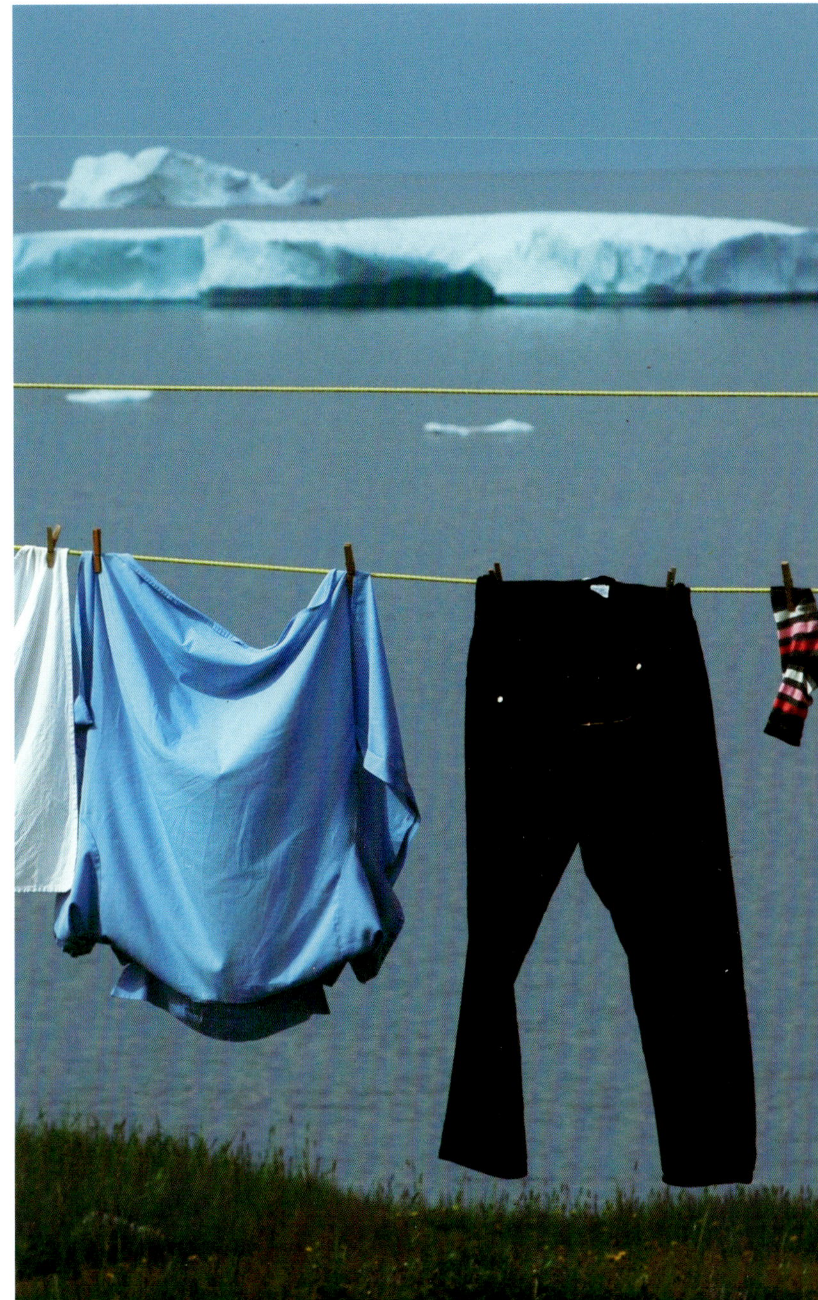

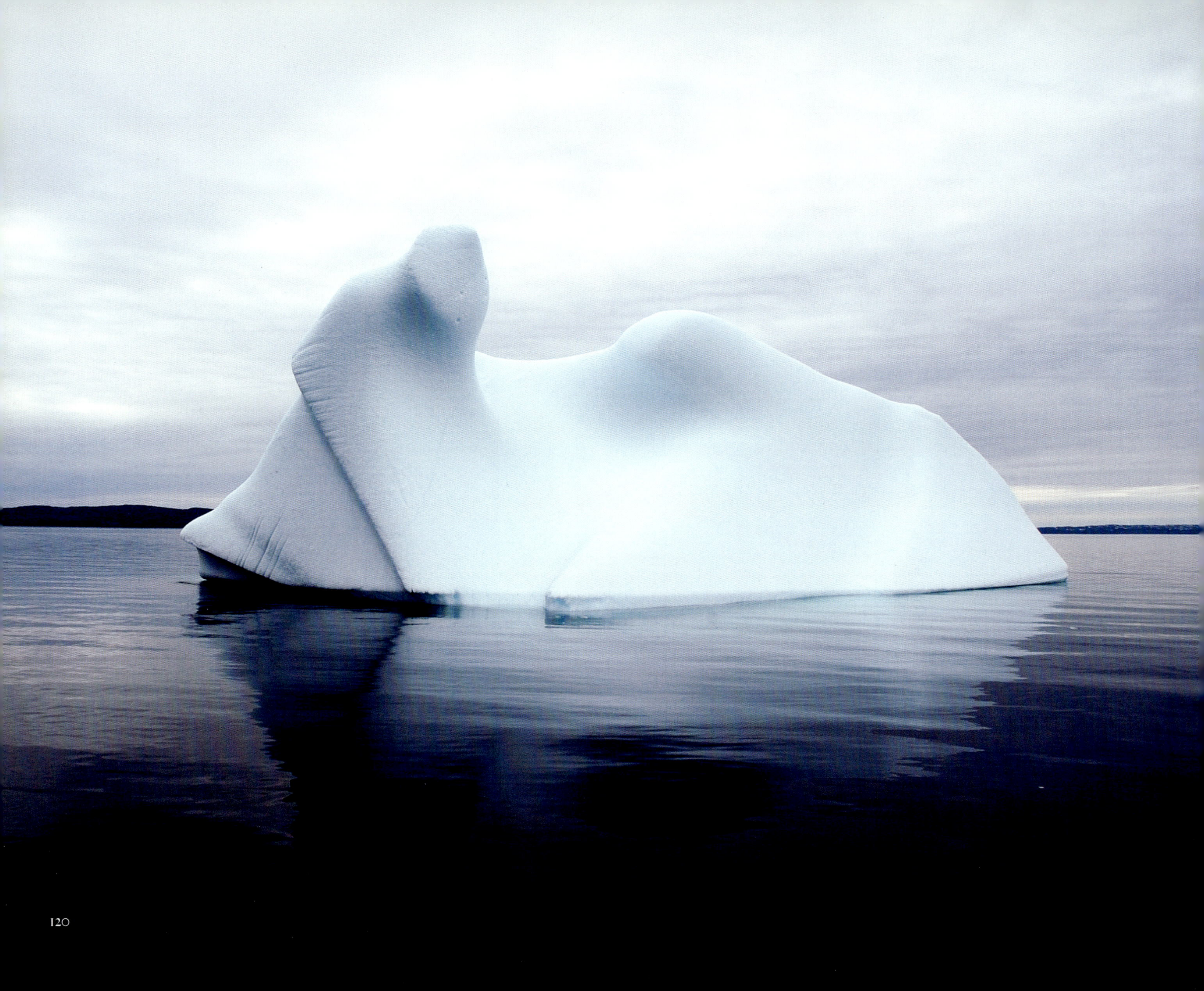

Old & Wrinkle-Free

Kevin

GREEN HARBOUR ISLAND

April 30, 8:55 a.m.

Bulges on the ocean's face reflected light lines parallel to the bay's shore . . . exfoliating like scales off a prehistoric sea monster. A sprinkling of mackerel danced in and out of the light attracting a forty-ton beast to the surface. With a *whoooooosh* the beast exhaled blowing a geyser of sea spray skyward. Like a submarine it settled at the surface, then, smacking the mackerel violently with its fluke, the fish were stunned into submission for the whale's morning meal. Holding its position as we approached, the whale fed voraciously. To our surprise, another whale appeared, mimicking the actions of the first humpback.

Our destination was a one-half-hour's steam to a relatively small 200,000-ton iceberg northeast of Green Harbour Island. The rich bait stock in the bay seduced the humpback whales to an early appearance . . . and now the whale was delightfully seducing us, delaying our voyage to the ice!

April 30, 10:40 a.m.

On arriving at the outskirts of Green Harbour, the bay's still surface reflected an ice-white silhouette. Like that of a giant seal lying on its belly with head perked upward and outward in our direction, we gazed back, pondering the creators of this beautiful shape. Smooth curvaceous lines marked the iceberg's edges like that of a beautifully sculpted woman. The paradox being it was over 10,000 years old, nearing the end of its life cycle, and remarkably, it was wrinkle-free!

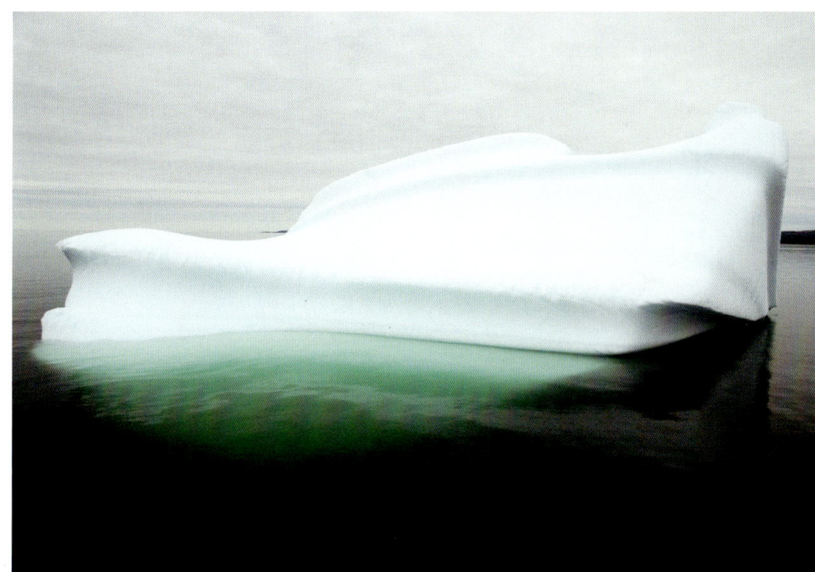

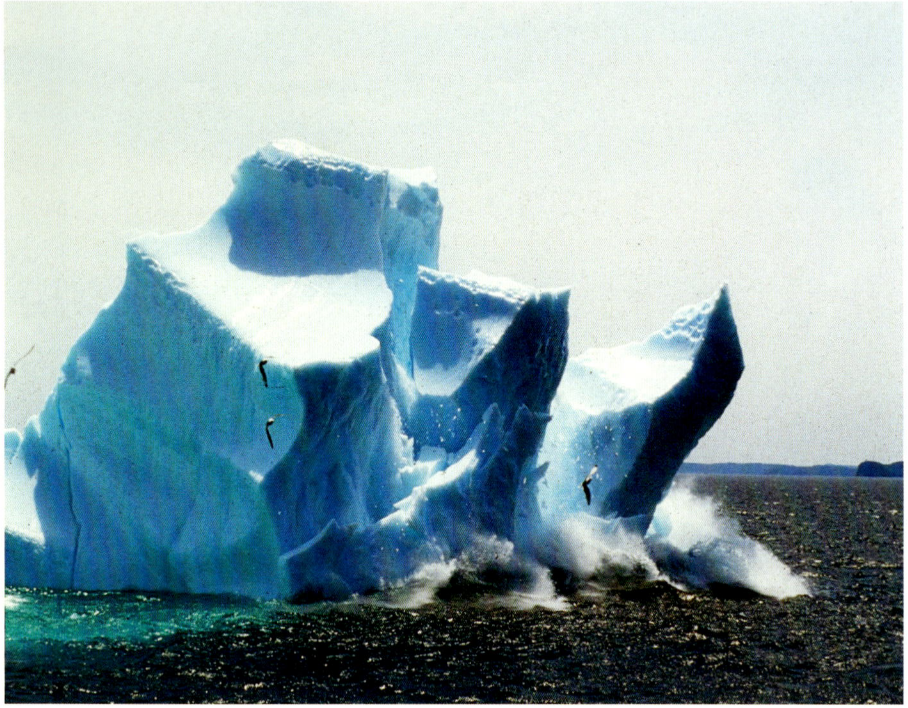

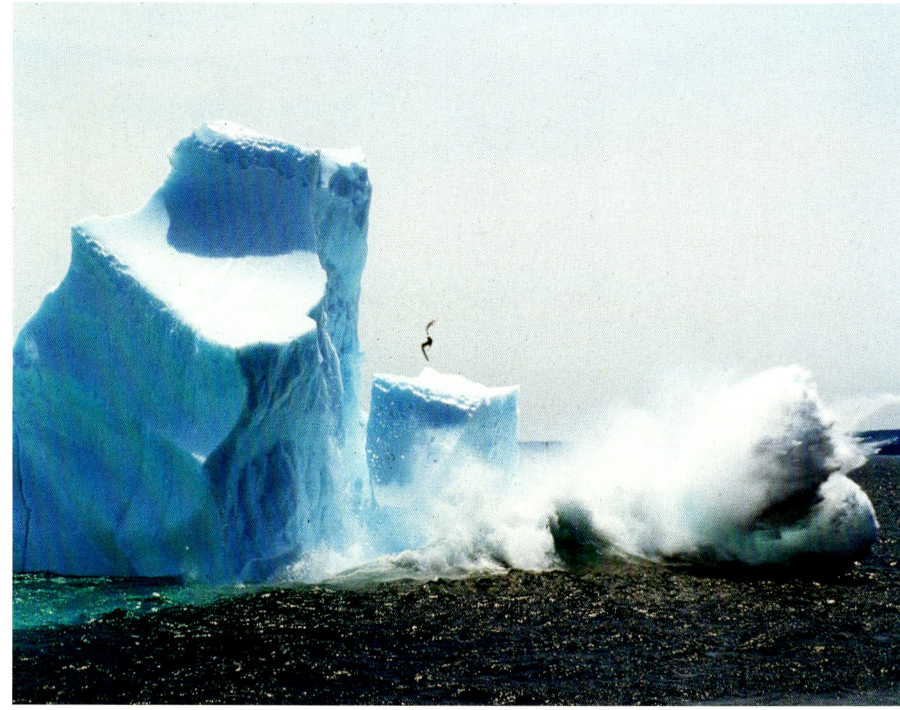

Collapse of an Iceberg

Kevin

Avalon: Witless Bay Seabird Sanctuary Photo Expedition

Collapse of an Iceberg

July 3 – 4, 1996

Some icebergs are remembered best for the surrounding events. The collapsing iceberg of Witless Bay is one!

It all began the day before. We landed on Gull Island, one of four islands included in the world-renowned Witless Bay Seabird Sanctuary. Here more than one million breeding seabirds, including ten different species, surrounded us.

The second-largest colony of puffins in the western North Atlantic is most obvious by day. The constant swirl of puffins leaving their burrows and returning with fish for their young embody an energy all its own . . . resembling a snowstorm in Technicolor. Dodging falling bird scat is an occupational hazard impossible to completely avoid.

The world's second-largest leach storm petrel colony, virtually invisible by day, return in waves to their breeding burrows at night; likely natural selection's defence mechanism to avoid these small fist-sized birds from being eaten by the voracious black-back gull and other predators. The eerie cacophony, elevated with ensuing darkness, seemingly lasting through the night . . . reminding us of Alfred Hitchcock's 1963 movie *The Birds* where birds of all kinds suddenly begin to attack people, in increasing numbers and with increasing viciousness. We wondered aloud: Was that our fate?

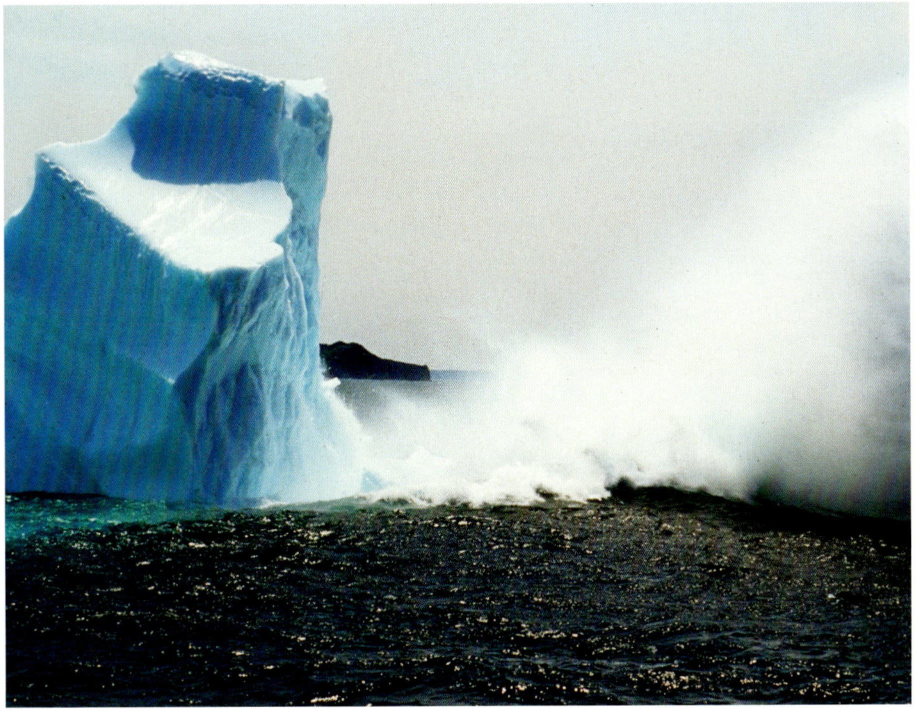

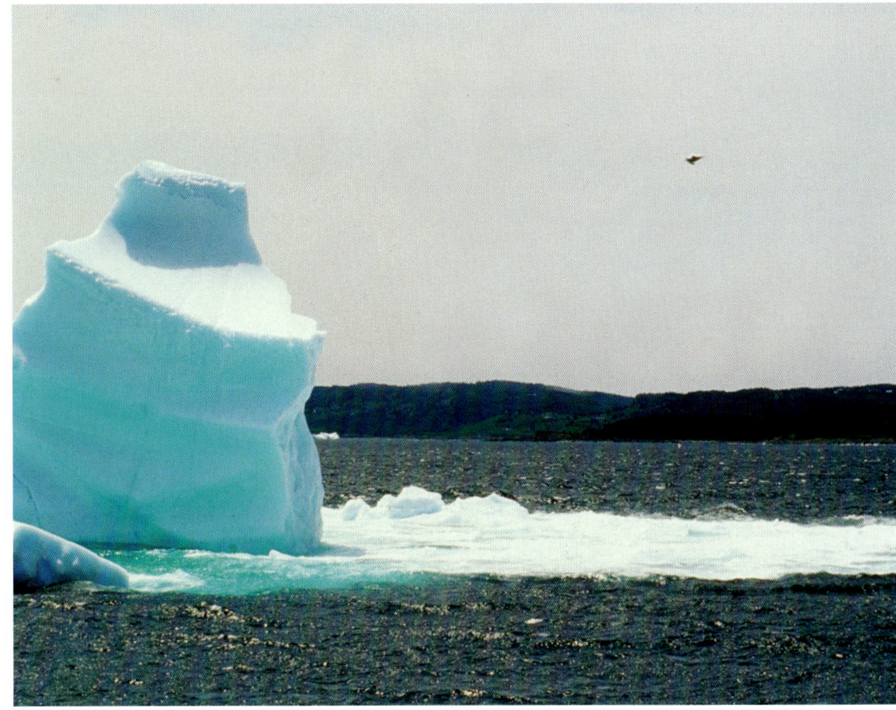

Our place of residence for the night, Gull Island, was a different place. Devoid of any rich vegetation except for grassy hummocks piqued with exposed peat burrows which are found on the steep slopes. In low-sloped higher areas, scattered dwarfed spruce and fir forest are conspicuously laden with bird scat and/or lichen, part of their own struggle to survive. This place, more than most places we have been, exemplifies survival of the fittest.

The bounty of fish attracting the birds also lures a variety of whales . . . most common being the minke and humpback. Today was no exception, with clusters of birds following these huge mammals scavenging leftovers. Our primary task was photographing the birds. Few permits are granted to landing on the island(s). The *National Geographic* film crew that left as we arrived, was a poignant reminder of privilege. The short list of photographers and film crews on the shack's wall from Japan, Germany, and Chile left us feeling a part of esteemed company.

The tour boat picking us up towed a small collector boat traditionally known as a "shallaway." As we and our gear were being collected and shuttled to the larger tour boat, we were the focus of every tourists camera. Then it was off to visit a mid-sized iceberg off Witless Point. I remember being on top of the boat, as high and unobstructed as possible with the crowd on board. Focused on the iceberg and in between shots, a thunderous clap emanated from the berg's core. As the berg split, in panic and hope I held the shutter release down like a panic button. The ensuing wave rocked the boat, I wobbled . . . my knees were shaking! Three days later, with knees shaking again, I looked at the freshly processed slides. Yes! I got the shot . . . a sequence of four shots as the iceberg collapsed. One of those shots transpired from a rendezvous of luck and fate.

P.S. *Next time in company of so many birds . . . bring wide-brim hat!*

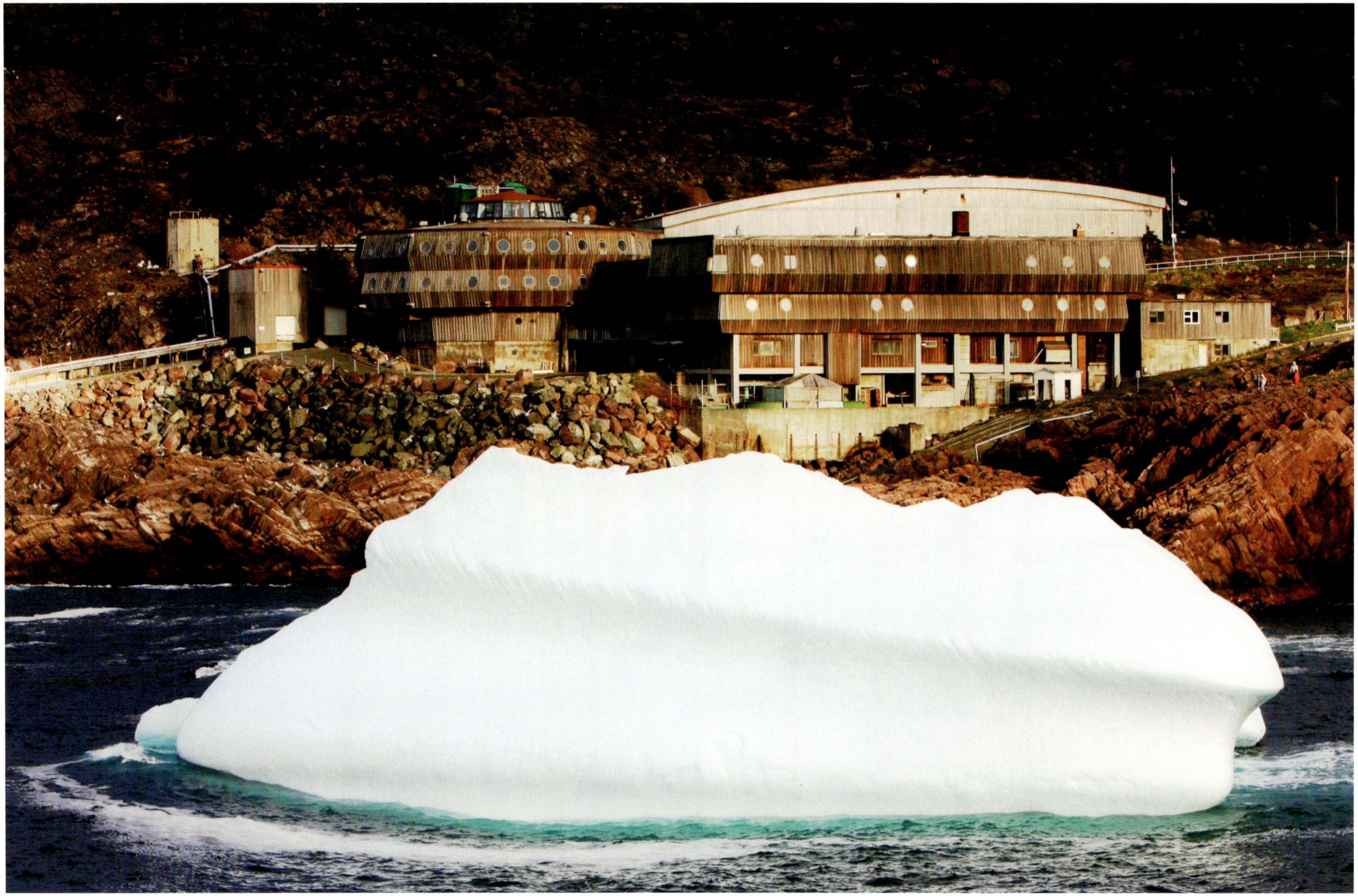

A Stone's Throw Away

Steve

Logy Bay

We often say that "icebergs return to the same spot every year." Now, we know an iceberg can't reappear after melting away, so it is fairly obvious that what is meant by "return" is the repeated presence of an iceberg in a particular spot year after year.

Of the many iceberg lairs I have watched over the years, Logy Bay is one of my favourites. The reasons are many: the ease and comfort with which to observe, the heavy swell and deep waters of the narrow inlet, the access to high and low ground practically all around, including interesting scenery and structures like Redcliff and the Marine Lab. Logy Bay is one of the few places where I have repeatedly listened to an iceberg crunching into the crags below water as large swells heave and drop the colossal mass. Oddly it is here that I often have the primal urge to throw a rock at the big moving thing, perhaps because it is a rare pleasure to be able to prove to oneself that you are actually within a stone's throw of something extraordinary.

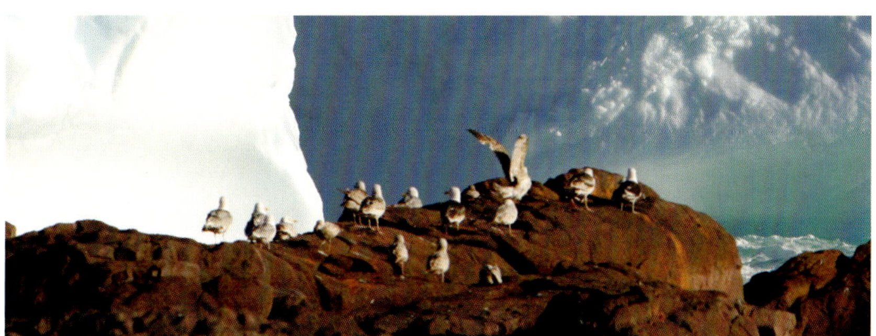

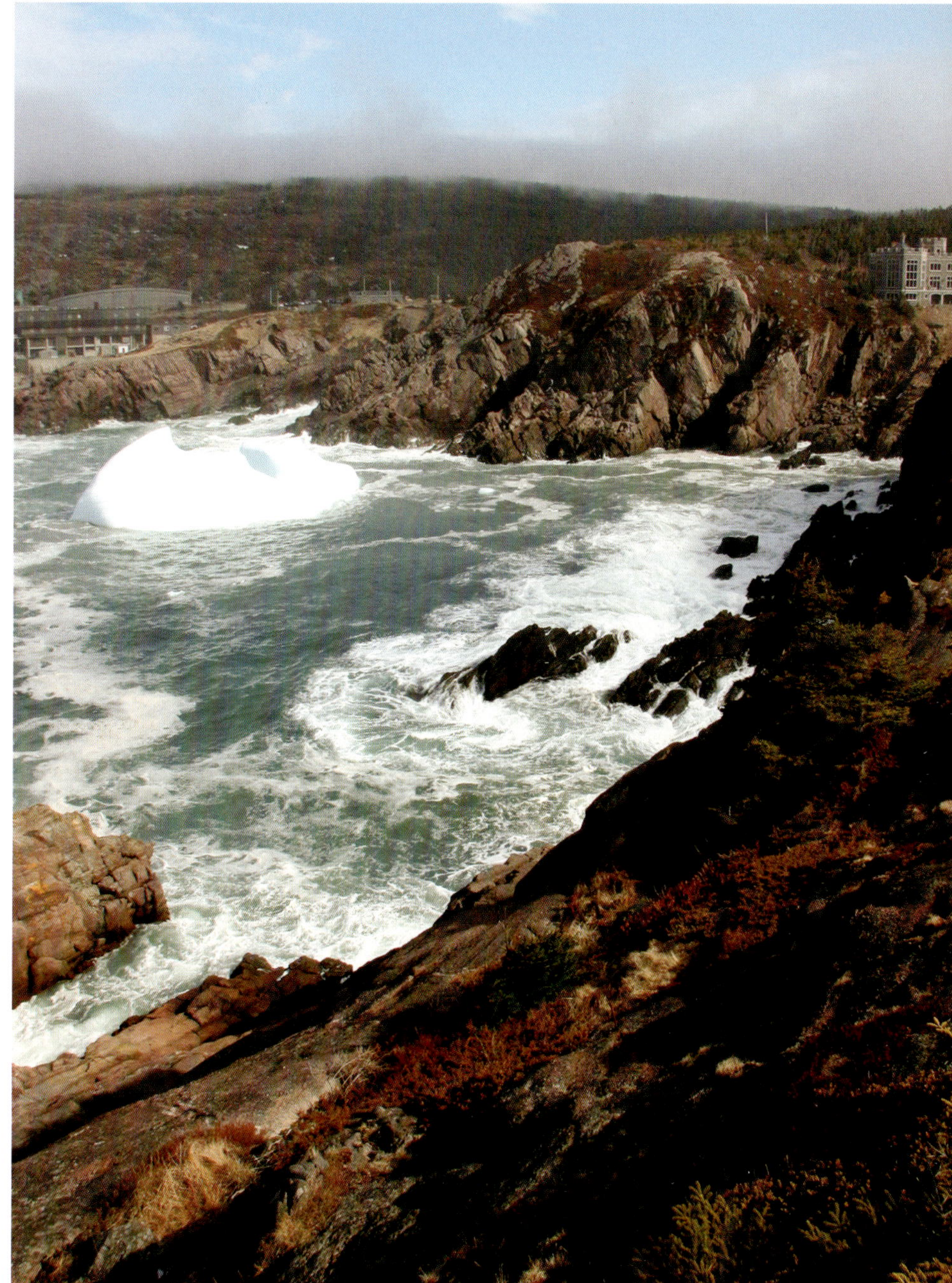

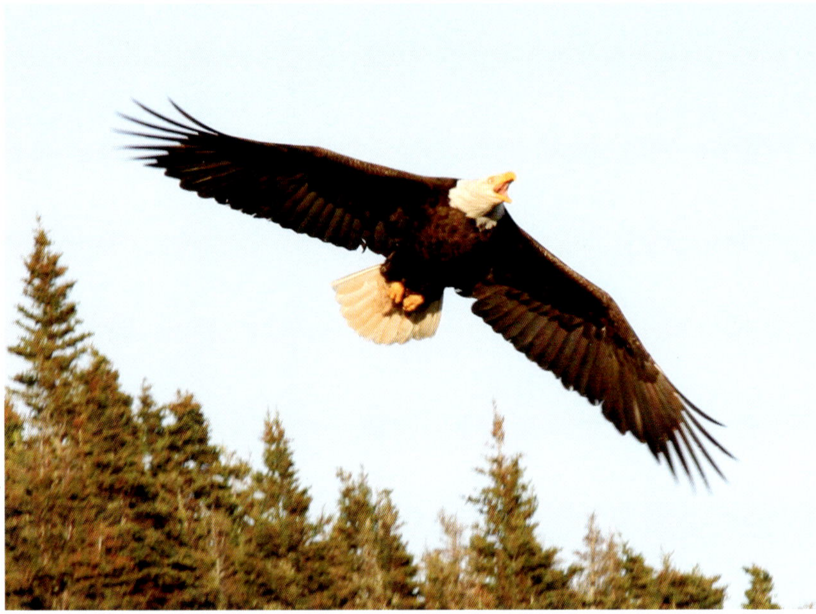

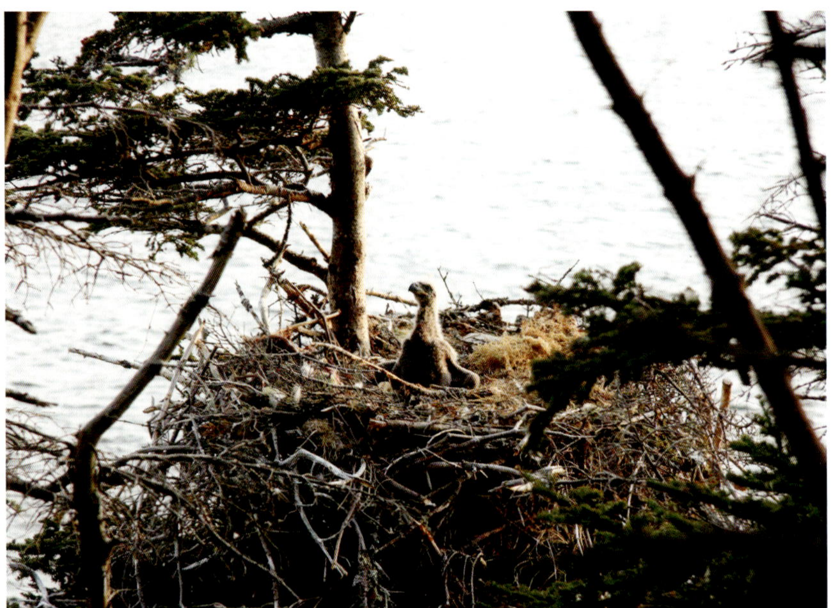

Herring Cove Eagles

Steve

I had been on a whale-watch tour in Bay Bulls earlier that day and decided to continue south, or "up" the Southern Shore for an evening drive. Rounding the corner in Mobile I saw a fantastic drydock iceberg very near shore only about one kilometre through the woods from the road. Conveniently, a piece of the East Coast Trail led directly to the spot where the berg was nestled, indicating it by name as Herring Cove. I was mesmerized by the looming berg as I emerged from the woods trail and stepped down the open gentle grassy slope leading to the shore. Watching my footing gave me no warning of what was to happen next – in an instant a shrill scream pierced my mind and a dark flash swept over me.

Seems that the iceberg had found its way into the lair of a nesting pair of bald eagles. Though I had clearly intruded, the raptors eventually concluded that I wasn't a threat and so they calmed and perched opposite the cove where I sat gathering my wits. I didn't know which way to turn or what camera lens to use; there was something to see in all directions and at all scales. It was an awesome sight that evening, to sit and watch the sun slowly set over this place where the paths of these awesome natural wonders briefly crossed.

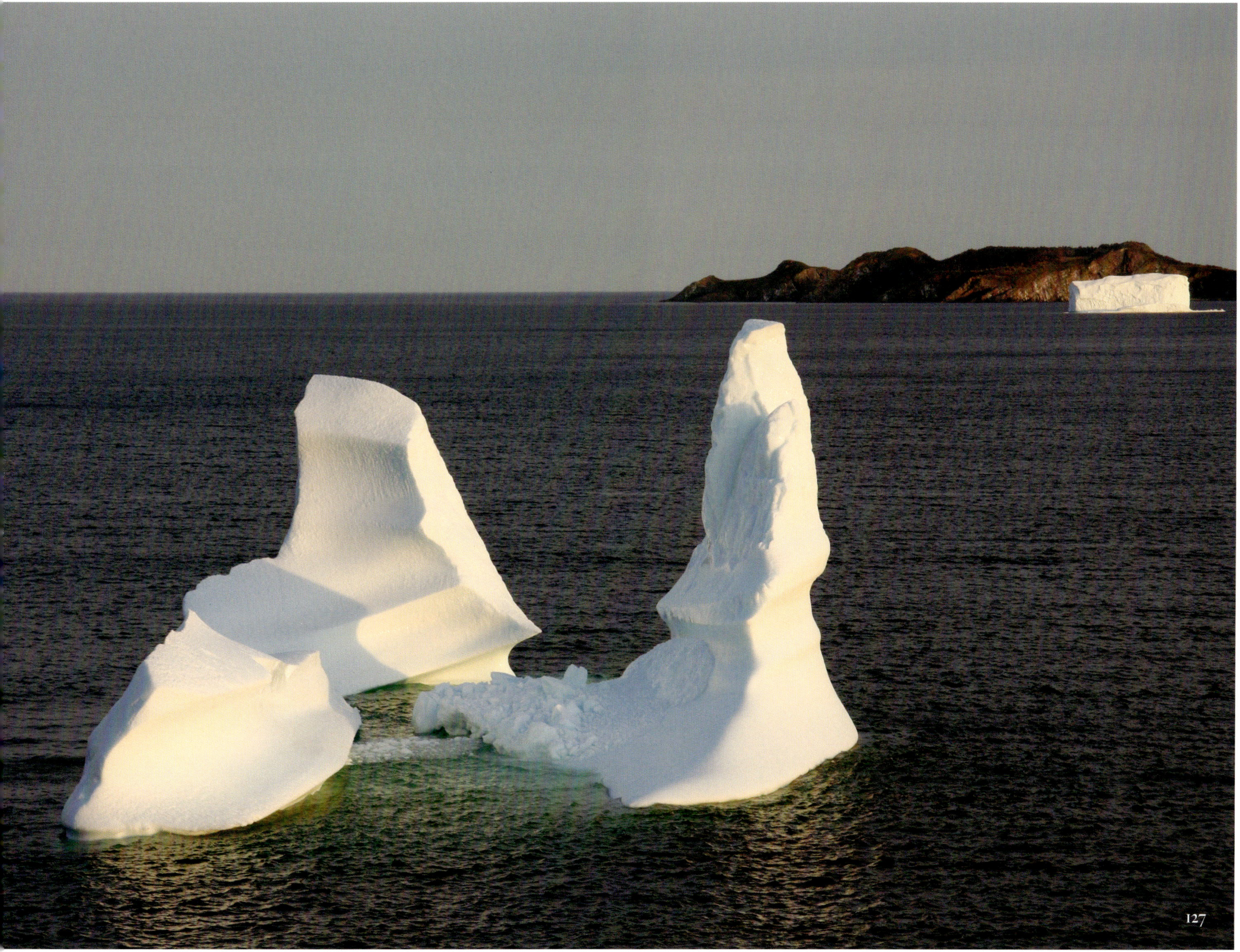

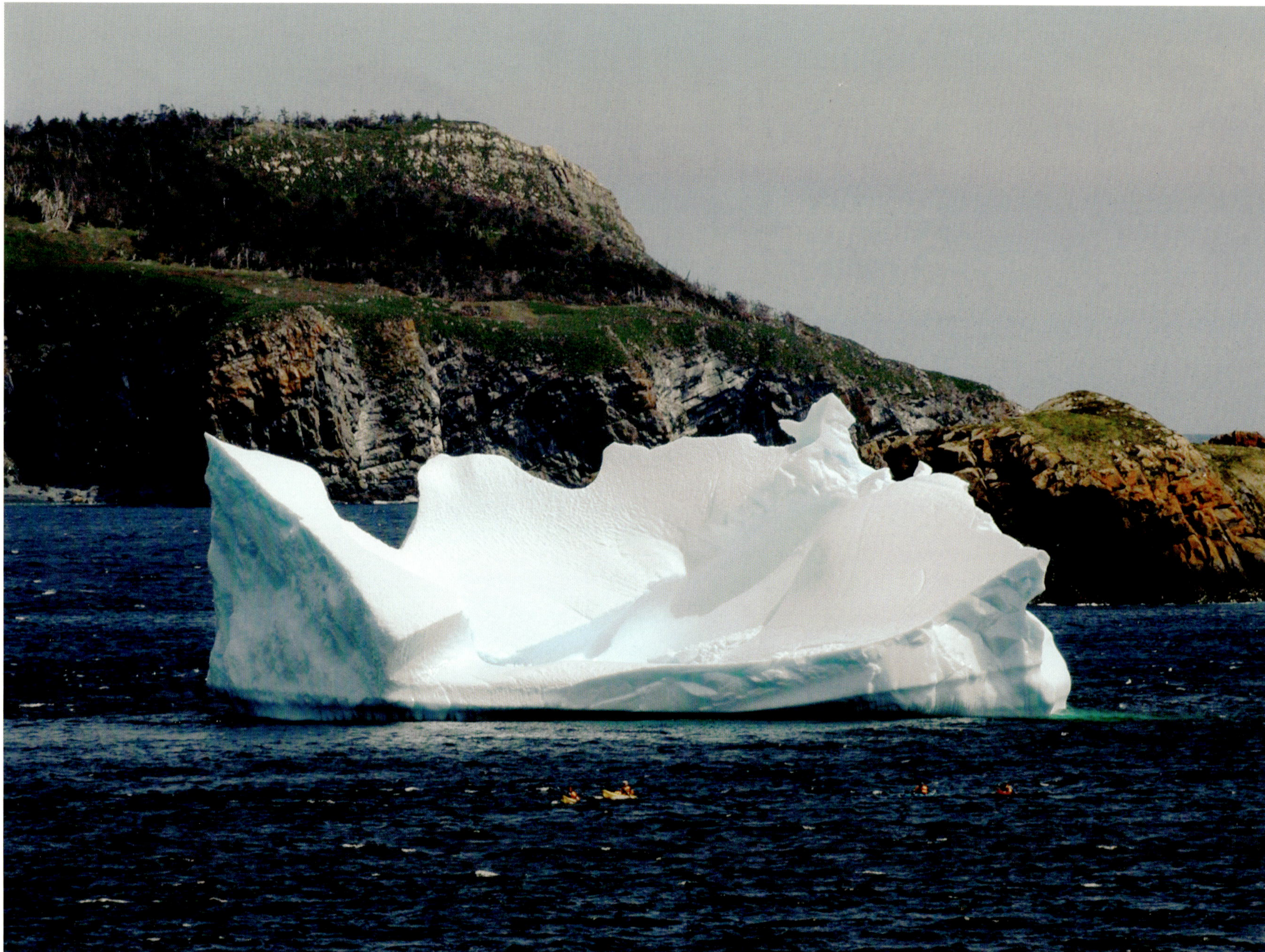

Adventuresome Tots

Kevin

JUNE 8, 2008

Tors Cove is a beautiful place at the worst of times. Yesterday, Saturday, had been a day of coastal fog and flat light, but it still brought a steady stream of optimists seeking a peek of the icebergs along the Avalon Peninsula's southern shore. The iceberg in Tors Cove was grounded off Ship Island. I took a few pics, but they were cursory at best . . . not what we were looking for. We needed better light and something for scale.

Today, taking my mom for a Sunday drive, our first stop was Tors Cove. Light was brighter than yesterday, winds appeared to be strengthening, seas were blue, dotted with a growing number of whitecaps; the iceberg hadn't moved since yesterday, so I knew right where to go for iceberg photographs from different perspectives in the community.

From a seaside hilltop I began to shoot the iceberg and island with the big zoom lens, when to my delight I saw a kayak, then another, then another two. They were paddling diagonally across the face at a rapid pace, responding to the strengthening wind with obvious gusto. The kayaks offered wonderful scale; here even I could now relate to the actual size of this iceberg; it was much bigger than I had perceived yesterday or before the kayaks entered the frame. Continuously shooting with the kayaks and iceberg in view, it wasn't long before the kayaks approached shore, revealing two doubles and two singles.

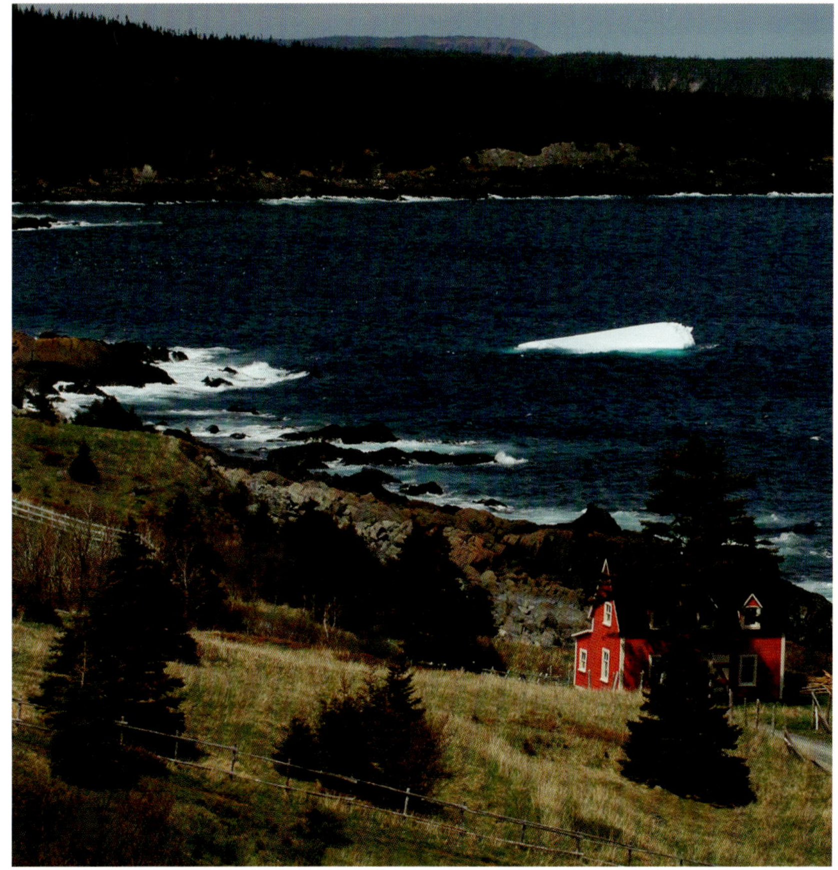

As the lead double approached, I laid my camera aside to help beach the lead boat. Interestingly, the bow paddler in both doubles was not paddling; in fact, they never even had a paddle. As I hauled the first double kayak ashore, the bow paddler comfortably stood, ejecting himself from the cockpit, and to my surprise he was maybe six or seven years old. In the second double kayak was his little sister, at four years old. All I could think was how lucky they were and what adventures they would be sharing on the front porch sitting in rocking chairs when they are my mother's age of ninty-five years young!

Signal Hill, Cape Spear & Beyond

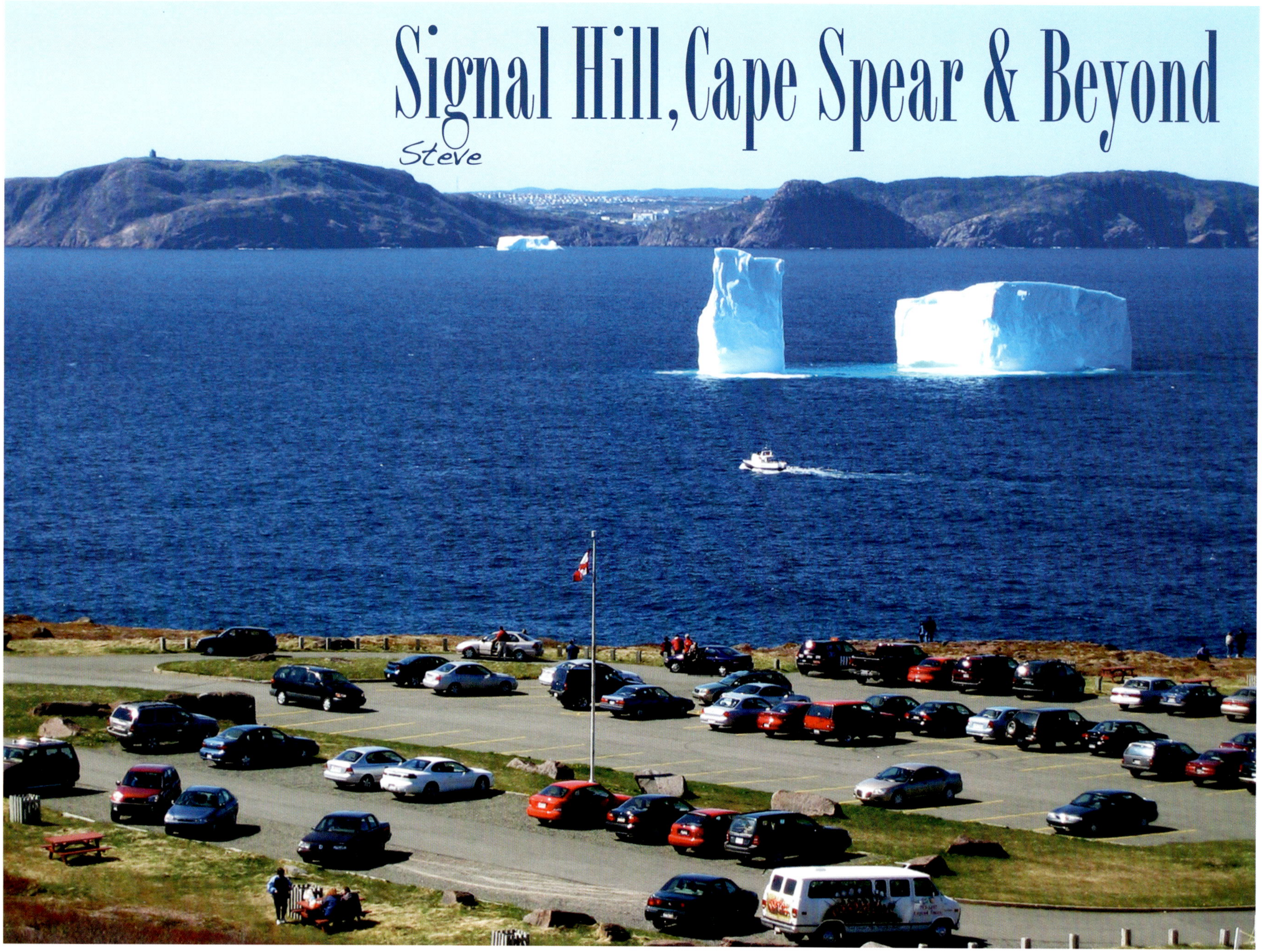

As a young kid I was familiar with icebergs because my father was involved in the earliest sponsored experiments to tow them. I remember my mother handing him a big can of peanuts before boarding the *Percy Crosby* and then later watching the expedition as it sailed out the narrows past Signal Hill and then Cape Spear, to some mysterious place beyond in search of giant icebergs. We usually had bergy chunks in our freezer at home and Sunday drives in the spring often involved scouting the coast for signs of their arrival. Still, I don't recall any special interest in icebergs – until in adulthood when I started taking note of the profound interest that people living elsewhere had in them. Later I would come to work as a research engineer and was involved in my own iceberg towing, crushing, melting, and modelling experiments – and collectively my experiences led others to ask for information and stories, a prompting which eventually led to the publication of the 2004 field guidebook entitled *Icebergs of Newfoundland and Labrador.* Though the field guide has served to inform, there remained in me an interest to produce a picture book which in some way might compliment the guide but, more importantly, would be an exposé of icebergs intended to communicate with pictures what words cannot capture.

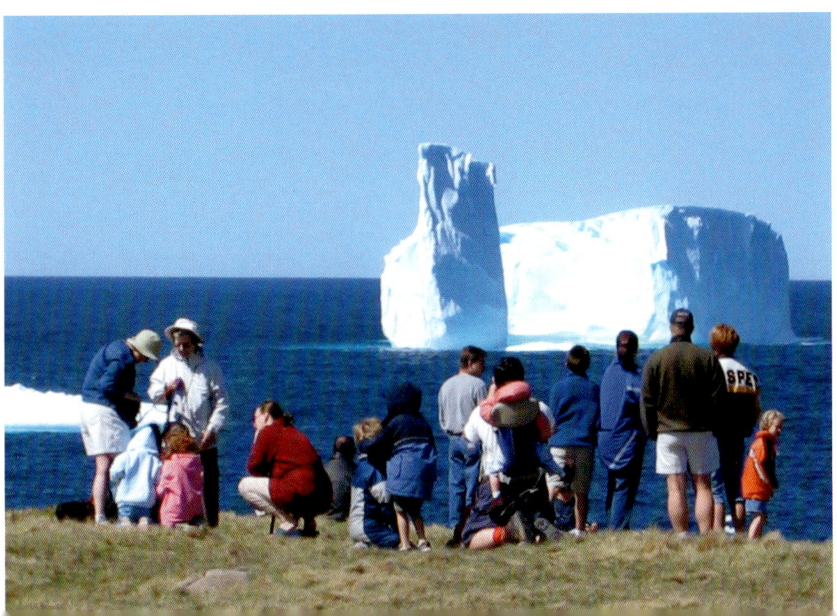

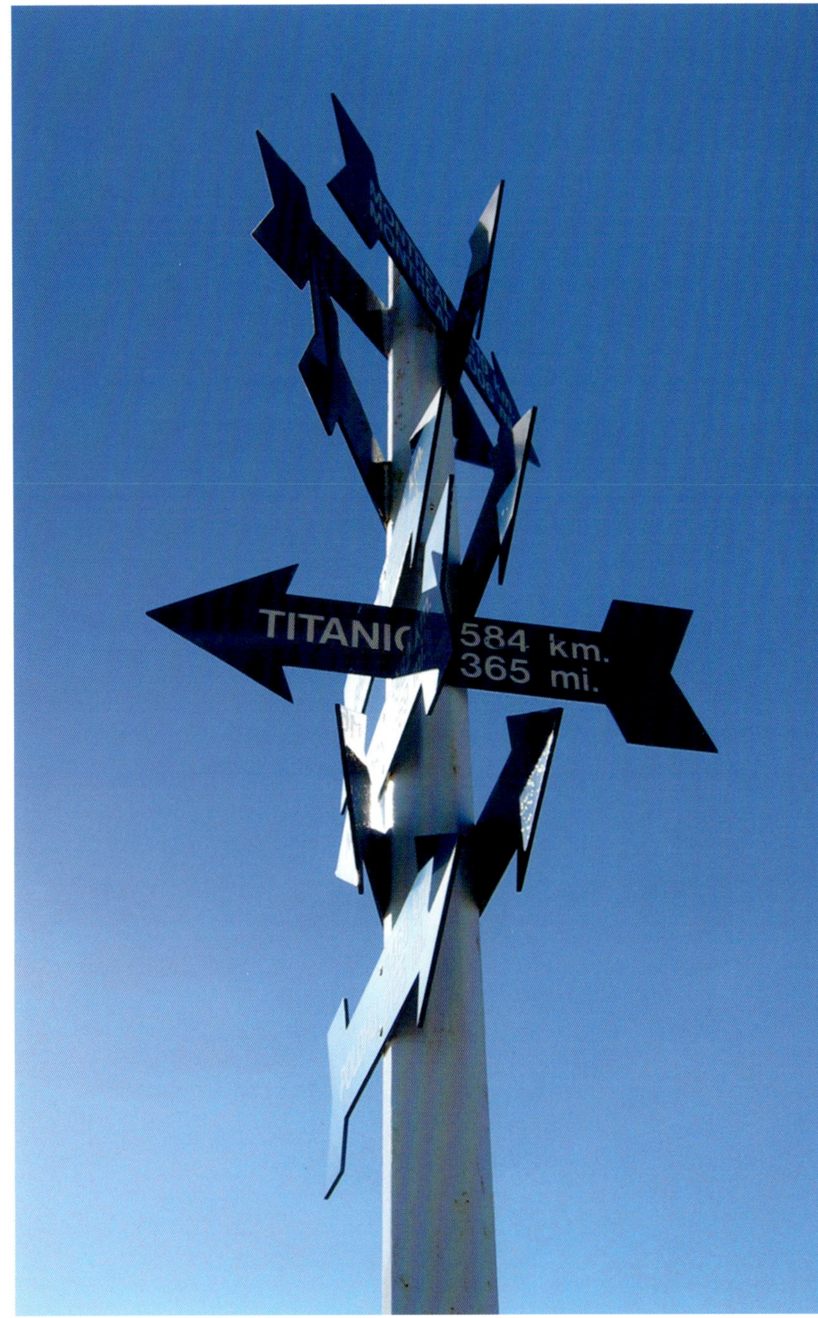

About five years ago I happened to bump into Kevin Redmond, with whom I was acquainted through prior tourism workshops and community involvement. Kevin, I knew, was an experienced photographer, and had published a few books on his considerable experience as a sea kayaker, explorer and outdoorsman. In an instant it occurred to me that Kevin might be interested in partnering on this iceberg picture book project, as his experience on the water, with cameras, and in publishers' offices would compliment my own skills neatly. He accepted immediately, and so began our joint quest to capture the perfect shot and share it with the world. What we couldn't know at that point was that we in Newfoundland were entering a severe iceberg drought that would persist through two seasons and would eventually drive us both to some ridiculous acts of photographic desperation.

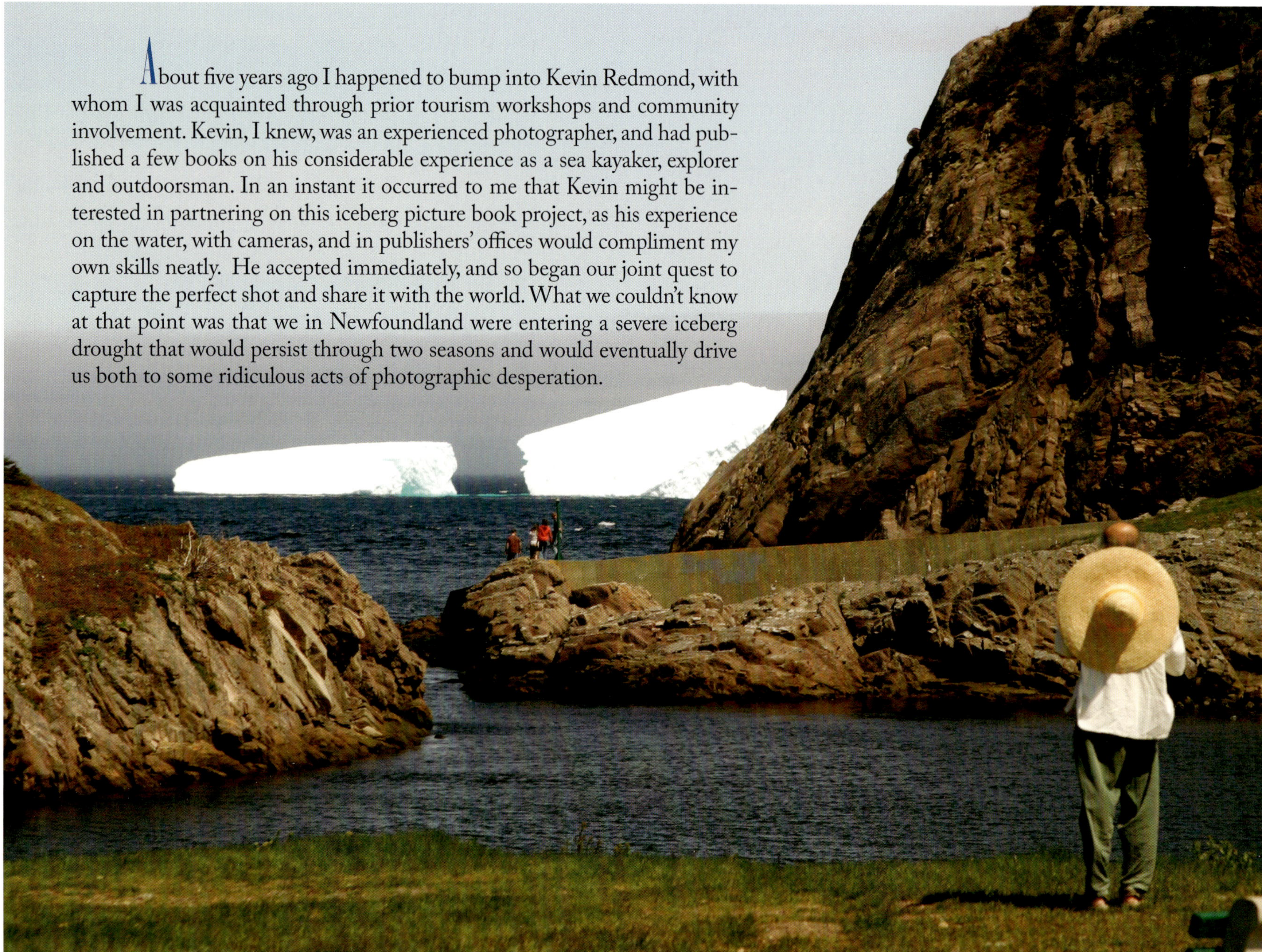

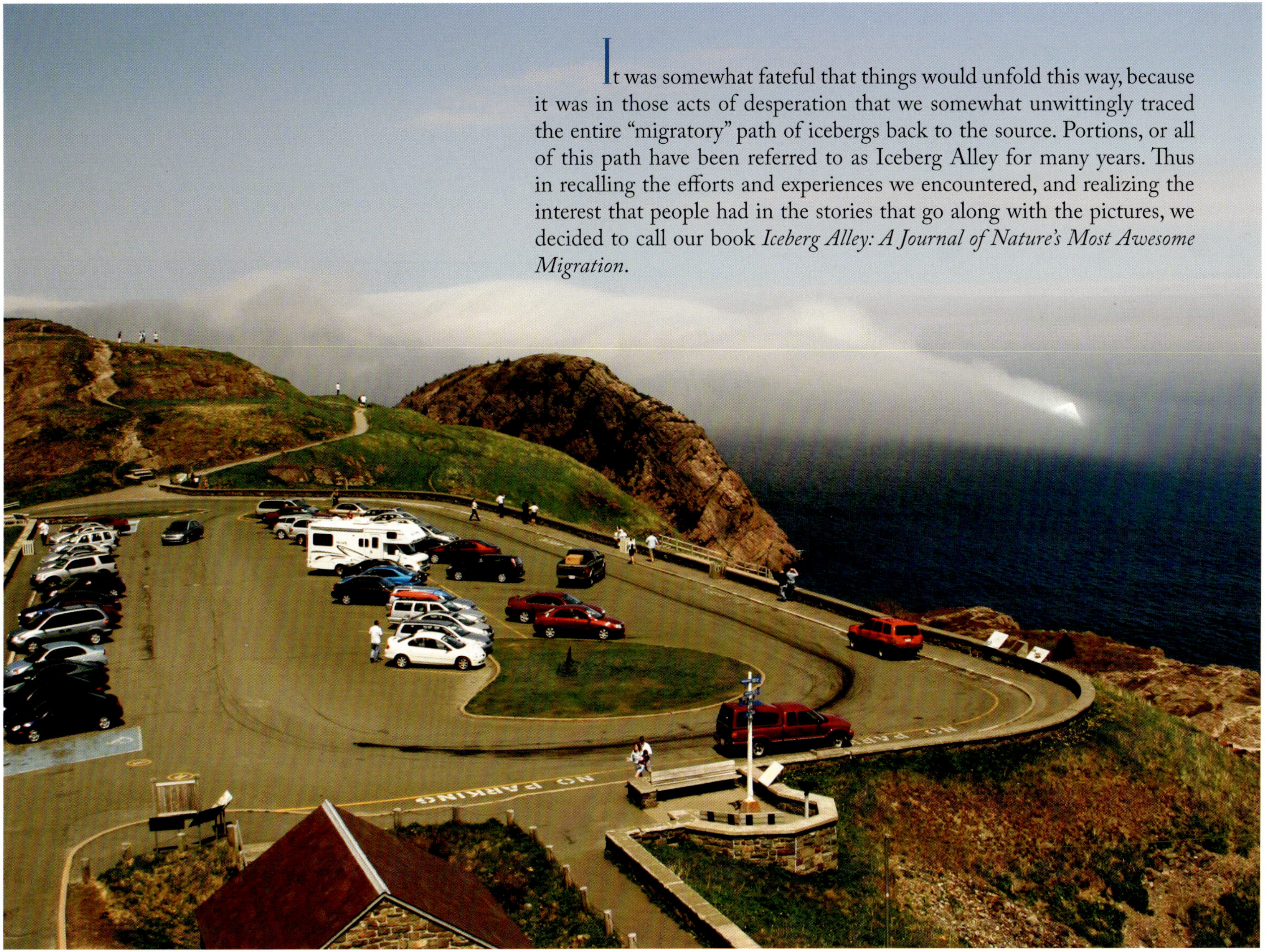

It was somewhat fateful that things would unfold this way, because it was in those acts of desperation that we somewhat unwittingly traced the entire "migratory" path of icebergs back to the source. Portions, or all of this path have been referred to as Iceberg Alley for many years. Thus in recalling the efforts and experiences we encountered, and realizing the interest that people had in the stories that go along with the pictures, we decided to call our book *Iceberg Alley: A Journal of Nature's Most Awesome Migration*.

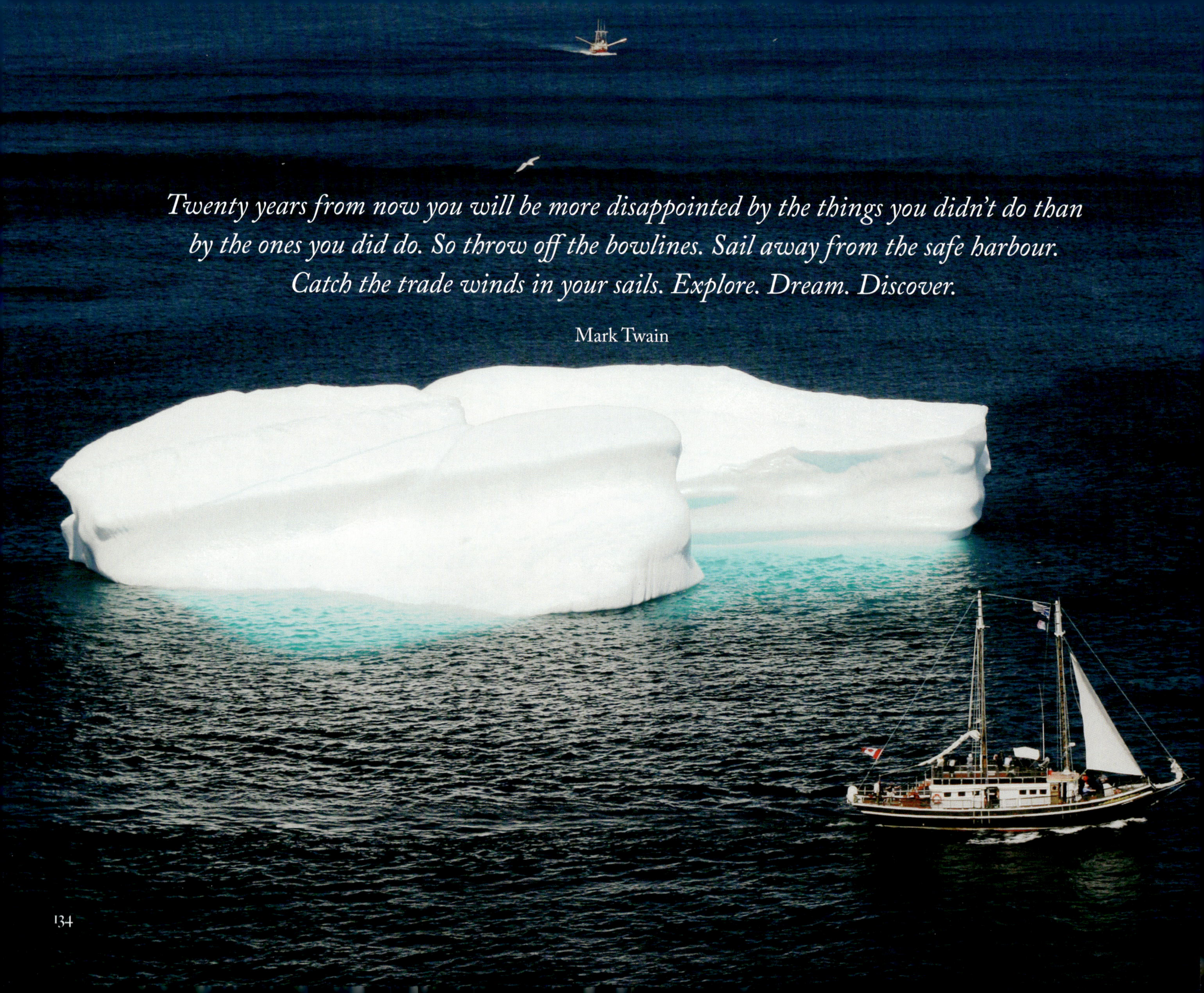

Twenty years from now you will be more disappointed by the things you didn't do than by the ones you did do. So throw off the bowlines. Sail away from the safe harbour. Catch the trade winds in your sails. Explore. Dream. Discover.

Mark Twain

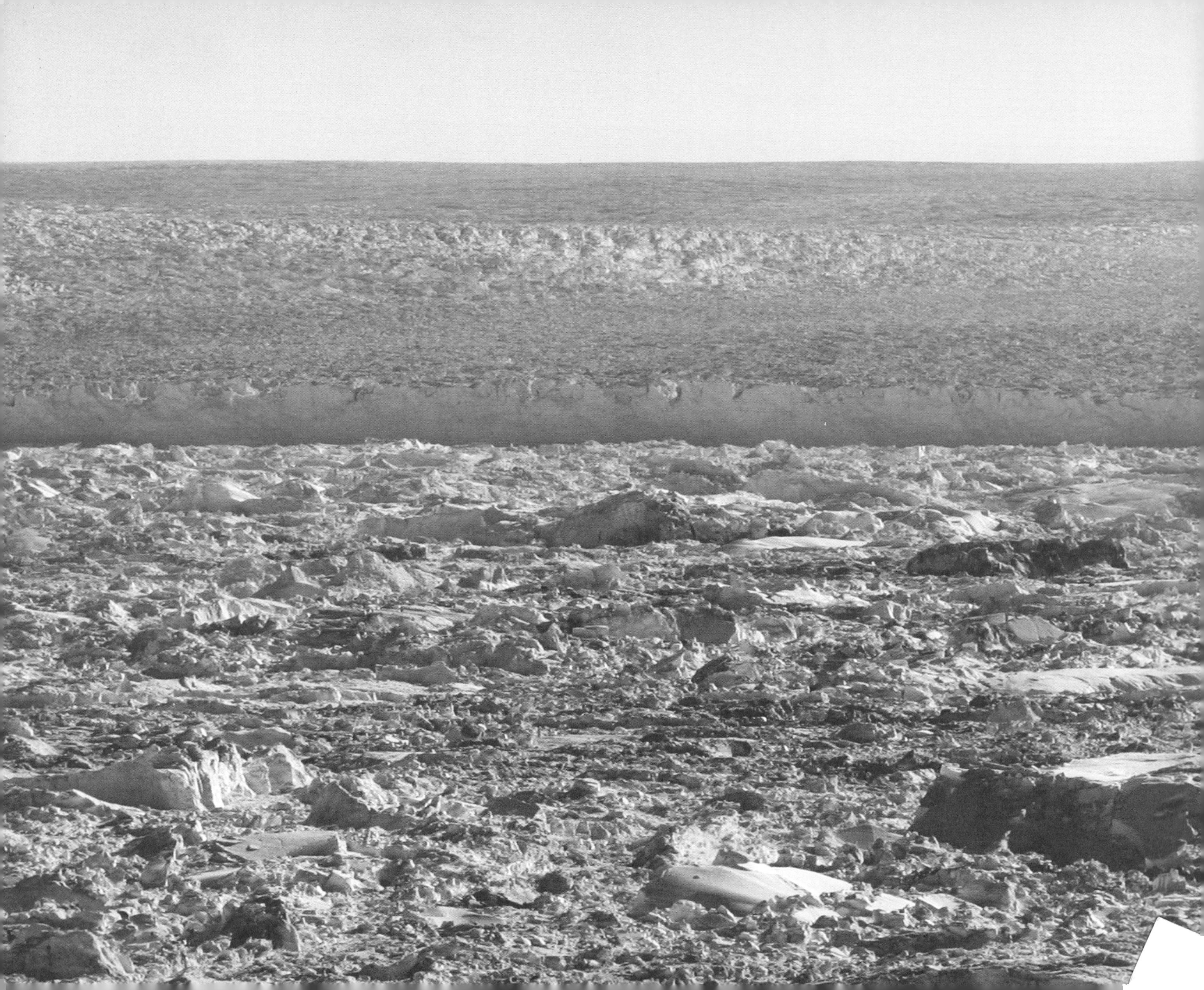